GENERALIZED CLASSICAL MECHANICS
AND FIELD THEORY

NORTH-HOLLAND MATHEMATICS STUDIES 112
Notas de Matemática (102)

NORTH-HOLLAND – AMSTERDAM • NEW YORK • OXFORD

GENERALIZED CLASSICAL MECHANICS AND FIELD THEORY

A Geometrical Approach of Lagrangian and Hamiltonian Formalisms Involving Higher Order Derivatives

Manuel DE LEÓN

Departamento de Geometria y Topologia
Facultad de Matemáticas
Universidad de Santiago de Compostela
España.

and

Paulo R. RODRIGUES

Departamento de Geometria
Instituto de Matemática
Universidade Federal Fluminense
Niterói, RJ
Brasil

1985

NORTH-HOLLAND – AMSTERDAM • NEW YORK • OXFORD

ISBN: 0 444 87753 3

Publishers:
ELSEVIER SCIENCE PUBLISHERS B.V.
P.O. Box 1991
1000 BZ Amsterdam
The Netherlands

Sole distributors for the U.S.A. and Canada:
ELSEVIER SCIENCE PUBLISHING COMPANY, INC.
52 Vanderbilt Avenue
New York, N.Y. 10017
U.S.A.

PRINTED IN THE NETHERLANDS

"Il est peut-être inutile de faire observer qu'il n'est point de Mathématiques "sans larmes" à l'usage exclusif des physiciens et que, si je me suis efforcé de choisir et coordonner celles des théories mathématiques qui peuvent être utiles aux physiciens, il m'était impossible de renoncer dans leur exposé à cette rigueur sans laquelle il n'est plus de science ni mathématique ni physique."

A. Lichnerowicz.

(Algèbre et Analyse Linéaires, Masson, Paris, (1948)).

TO

Prof. Constantino M. de Barros†

Prof. Nelson L. Teixeira

Prof. Enrique Vidal Abascal

GENERALIZED CLASSICAL MECHANICS AND FIELD THEORY

FOREWORD

The aim of this book is to build up a large panel of the present situation of Lagrangian and Hamiltonian formalisms involving higher order derivatives. The achievements of Differential Geometry in formulating a more modern and powerful treatment of these theories are developed, including the contributions of the author's themselves. An extensive review of the development of these theories in classical language is also given.

A Lagrangian formalism is said to be of <u>higher order derivatives</u> if it is described by a real (smooth) function L which depends on n-independent variables x_a, m-functions $y^A(x_a)$ and all derivatives of the y 's with respect to the x's up to a certain finite order k. For sake of simplicity, we will say that L is a <u>Lagrangian of order</u> k. In Particle Mechanics and Field Theories one usually works with Lagrangians of order one. Therefore, higher order Particle Mechanics (resp. Field Theories) means that the Lagrangians depend not only on position/velocity variables (resp. independent coordinates/ field variables) but also on their time derivatives up to k-th order (resp. partial derivatives up to k-th order of the field variables with respect to the independent coordinates).

Higher order Hamiltonian formalisms will understand, as

in the standard theory of order one, the Hamiltonian counterpart of such Lagrangians.

There is not much agreement in the literature as to the interest in this kind of problem. It seems to have started with M. Ostrogradsky in 1848 (see Whittaker (1959)). According to P. Dedecker (1979), it was J. Jacobi who first studied these systems. Also Todhunter, in his book "History of the Calculus of Variations", mentions that Clebsch, following a suggestion of Jacobi, considered this subject in 1858. Therefore, we may call such theory Jacobi-Ostrogradsky Generalized Classical Mechanics (and Field Theory) or, more simply, Genegeneralized Classical Mechanics (and Field Theory).

In the last 40 years many papers dealing with higher derivatives in Mechanics and Field Theories have appeared. It seems that it was F. Bopp (1940) and B. Podolsky (1942) who renewed the interest in this kind of generalization in physics. Podolsky (and co-workers), for example, introduced an electromagnetic theory with second order derivatives.

Mechanics lays naturally in Differential Geometry and reciprocally. This "marriage" has allowed not only a more rigorous formulation from the mathematical point of view, but also a better understanding of its physical content. Following the results in Symplectic Mechanics systematized in the literature (see the "Bibles": Abraham & Marsden (1978), Arnol'd (1974) and Godbillon (1969)) it is found that a Lagrangian, resp. Hamiltonian, formalism can be characterized by geometric structures canonically associated to the tangent, resp. cotangent, bundle of a given differentiable manifold (they are respectively, the velocity, phase and configuration spaces). Analytical Mechanics is developed through geometric formalisms underlying the theory of fiber bundles.

These last years a certain number of papers where the geometrical formulation of this so-called Generalized Mechanics and Field Theory is developed have been published. Inspired by them, the major task of the present work will be to give an overview of the results obtained in the elaboration of this formalism, including our own results on the subject. The "natural place" of our study will be the Jet bundles, first introduced by Ch. Ehresmann in the years of 1950. For instance, in Lagrangian Particle Mechanics we develop the formalism on tangent bundles of higher order. We will emphasize some geometric structures underlying such a Mechanics in the sense that they are a generalization of the methods usually employed in the standard situation.

The text is divised in three chapters, each of them with an introduction. In the first we give some geometric tools necessary for the development of the others two chapters. In Chapter II we adopt the point of view of J. Klein (1962) for Lagrangian Mechanical Systems. Klein's Lagrangian formalism is developed with the help of the Almost Tangent Geometry (introduced by Clark & Bruckheimer in 1960) and a special exterior differential calculus. We extend this geometry to higher order tangent bundles. One advantage of such a choice is that it is possible to give an intrinsical exposition without carrying the symplectic form of the cotangent bundle, as we usually do for the standard regular Lagrangians. Therefore, we work, generically, with <u>pre</u>-symplectic structures in the place of symplectic ones. The third chapter is devoted to a local and sometimes global study of Generalized Classical Field Theory from the variational approach. The geometric formalism adopted there is the usual one underlying exterior differential calculus on manifolds.

This book is addressed mainly to graduate students. Of course it is assumed that the readers are acquainted with the geometrical formulation of standard Classical Mechanics.

We acknowledge the Brazilian Agency Conselho Nacional de Desenvolvimento Científico e Tecnológico (CNPq), Ministério de Educación y Ciencia, Spain, and the universities Federal Fluminense and Santiago de Compostela for their financial assitance during the preparation of the manuscript. Our thanks to the colleagues of our Departments, in special to Celso Costa and Jose Antonio Oubiña. To Ligia Rodrigues for her patience in reading the manuscript, pointing out some ambiguities and unclear points in the text. Finally, we would like to express our thanks to the editor of "Notas de Matemática", Professor Leopoldo Nachbin and the Mathematics editor of North-Holland, Dr. Arjen Sevenster, for including this volume in their series.

The authors.

TABLE OF CONTENTS

CHAPTER I

The Differential Geometry of Higher Order Jets and Tangent Bundles.

CHAPTER II

Generalized Classical Mechanics

CHAPTER III

Generalized Field Theory

CHAPTER I

THE DIFFERENTIAL GEOMETRY OF HIGHER ORDER JETS AND TANGENT BUNDLES

I.1. - Introduction

This chapter is devoted to the study of basic geometrical notions required for the development of the main object of the text. Some facts about Jet theory are reviewed in Section 2, and the reader may also consult the articles of Ehresmann (1951), (1954), (1955)(a), (1955)(b), the appendix of Aldaya & Azcárraga (1980), as well as the books of Golubitsky & Guillemin (1973) and Michor (1980).

In Section 3, a particular case of Jet manifolds is considered: the tangent bundle of higher order. We show that this jet bundle posseses in a canonical way a certain kind of geometric structure, the so called almost tangent structure of higher order, introduced by Eliopoulos (1966), and which is a generalization of the almost tangent geometry of the tangent bundle. This almost tangent geometry of higher order provides a special differential calculus which is a generalization of the formalism presented in the last chapters of the book of Godbillon (1969) on Differential Geometry and Classical Mechanics.

Another important fact examined in this chapter is the extension of the notion of "spray" to higher order tangent bundles. This concept was introduced for the ordinary situation by Ambrose, Palais & Singer (1960), and it is relevant in Mechanics. The theory of sprays is closely related to the theory of connections on manifolds; therefore, connections of higher order are introduced and the relation between sprays and connections is studied in some detail.

I.2. - Jet manifolds

2.1. Jets of sections

It is assumed throughout the text that all structures, mappings, etc., are smooth (C^∞-class). Let N and M be manifolds with $\dim N = n$.

DEFINITION (1). Consider the triple (M,p,N), where $p\colon M \to N$ is a mapping. We say that M is a fibered manifold over N with projection p if the following conditions are verified:

(i) $\dim M = n+m$, where m is a positive integer,

(ii) p is a surjective submersion,

(iii) for any point in M there are two charts (U,f) and (V,g) of M and N, respectively, with $p(U) = V$,

such that $p_1 \circ f = g \circ p$, where $p_1\colon \mathbb{R}^n \times \mathbb{R}^m \to \mathbb{R}^n$ is the canonical projection onto $\mathbb{R}^n$.

REMARK (1). We can easily check that conditions (i) and (ii) are a direct consequence of condition (iii).

A mapping $s\colon N \to M$ is said to be a section of M if $p \circ s = \mathrm{Id}_N$ (identity on N) and is said a local section if $p \circ s/U = \mathrm{Id}_U$, for an open subset U of N. We put $\mathrm{Sec}(M)$ for the set of all sections on M and $\mathrm{Sec}_U(M)$ for local sections (sometimes we say "s is a section along U").

Let (M,p,N) be a fibered manifold and F a real vector space of finite dimension. Suppose that

(i) for each $a \in N$, the fiber $N_a = p^{-1}(a)$ admits a vector space structure isomorphic to F;

(ii) there is an open covering $\{U_i\}_{i \in I}$ of N so that for each $i \in I$ there exists a diffeomorphism

$f_i: p^{-1}(U_i) \to U_i \times F$ such that for each $a \in U_i$ the restriction $f_i|_{N_a}$ is an isomorphism from N_a to $\{a\} \times F$; then (M,p,N) is said to be a (locally trivial) <u>fibered vector bundle</u>, or, simply, vector bundle (for more indications on vector bundles see Appendix A).

If $p: M \to N$ is a fibered manifold we shall denote by $\mathrm{Ver}\, M$ the <u>vertical bundle</u> of M over N, that is, $\mathrm{Ver}\, M$ is the vector subbundle of TM consisting of all tangent vectors to M which are projected onto 0 by Tp.

DEFINITION (2). Way say that $s, s' \in \mathrm{Sec}(M)$ are $\sim_k$-<u>related</u>, $0 \le k \le \infty$, in a point $x \in N$, if

(i) $s(x) = s'(x)$;

(ii) for all functions $f: M \to \mathbb{R}$, the function $f \circ s - f \circ s'$: $N \to \mathbb{R}$ is "flat" of order k at x, that is, this function and all their derivatives up to order k, included, are zero at x.

DEFINITION (3). The equivalence class determined by $\sim_k$ is called <u>jet of order</u> k, or, simply, k-<u>jet</u> at x. For $s \in \mathrm{Sec}(M)$, the k-jet of s at x is represented by $j^k s(x)$,

$j^k_x s$ or $\tilde{s}^k(x)$. The set of all k-jets at x is denoted by $J^k_x(M,p,N)$. We put $J^k(M,p,N)$ for the union $\bigcup_x J^k_x(M,p,N)$. When the fibered manifold is the same throughout the text we put simply $J^k M$ or, else, J^k.

It is possible to define jets for local sections. In such a case, it is simpler to work with germs of sections. A <u>germ</u> of a section is the equivalence class determined by the relation: two sections are related if they have the same value at every point in the intersection of their domains.

2.2. Jets of mappings

Let N and M be manifolds. In a similar way as we did for sections, we may consider a more general situation defining the notion of k-jet at a point for a mapping from N to M. If $f\colon N \to M$, then the equivalence class determined by $\sim_k$ is called the k-jet of f at x. We put also $j^k f(x)$, $j^k_x f$ or $\tilde{f}^k(x)$ for a representative of the class. The set of k-jets at x is now represented by $J^k_x(N,M)$ and, also, $J^k(N,M) = \bigcup_x J^k_x(N,M)$. In a similar way, we can define k-jets for mappings defined locally on N (and so, we may consider k-jets of their germs).

REMARK (2). We say that a fibered manifold (M,p,N) is <u>trivial</u> if there is a manifold B such that $M = N\times B$. Since the graph of any map $f\colon N \to B$ is the corresponding section of the fibered manifold $N\times B$ over N, we can speak equivalently of a map from N to B or of a section of $N\times B$ over N. Throughout the text we will consider only this situation and we shall dentify $J^k(N\times B,p,N)$ with $J^k(N,B)$.

It can be shown that the set of all k-jets of sections (or maps) admits a (smooth in our case) differentiable structure (see Pommaret (1978) or Golubitsky & Guillemin (1973), for example).

The k-jet manifold of mappings (or sections) can be fibered in different ways:

$\alpha^k: J^k(N,M) \to N$; $\alpha^k(\tilde{f}^k(x)) = x$ (called <u>source</u> projection)

$\beta^k: J^k(N,M) \to M$; $\beta^k(\tilde{f}^k(x)) = f(x)$ (called <u>target</u> projection)

$\rho^k_r: J^k(N,M) \to J^r(N,M)$; $\rho^k_r(\tilde{f}^k(x)) = \tilde{f}^r(x)$, where $r \leq k$.

The manifold $J^k(M,p,N)$ is a submanifold of $J^k(N,M)$ and so the above projections admit restrictions to it (denoted also by the same symbol). $(J^k(M,p,N), \alpha^k, N)$, $(J^k(M,p,N), \beta^k, M)$, etc., are the corresponding fibered manifolds. It is clear that $\alpha^k = p\circ\beta^k$ and $\alpha^r \circ \rho^k_r = \alpha^k$, where $r \leq k$.

2.3. The k-jet prolongation

Let N and M be manifolds. (M may be fibered over N). Let J^kM be the manifold of k-jets (for sake of symplicity, we identify here both notations).

DEFINITION (4). The mapping which associates every point $x \in N$ to the k-jet of a mapping $g: N \to M$ at x is called the k-<u>jet prolongation</u> (or <u>extension</u>) of g and is represented by $j^k g$ or $\tilde{g}^k$.

So $\tilde{g}^k: N \to J^kM$ is defined by $x \to \tilde{g}^k(x)$. It is clear that $\tilde{g}^k$ is a section of the fibered manifold (J^kM, α^k, N)

(the same considerations are true for the local situation). It is also clear that $\beta^k \circ \tilde{g}^k = g$.

Let us consider the fibered manifolds (M,p,N) and $(J^k M,\alpha^k,N)$. If $u: N \to J^k M$ is a section, then, in general, only locally there exists a section $s: N \to M$ such that $\tilde{s}^k = u$.

2.4. A particular situation

Let $N = \mathbb{R}^n$ and $M = \mathbb{R}^n \times \mathbb{R}^m$. We identify the sections of such (trivial) manifold to the mappings from $\mathbb{R}^n$ to $\mathbb{R}^m$ as well as their k-jets. If $f: \mathbb{R}^n \to \mathbb{R}^m$ is represented by

$$f(x) = (f^1(x_1,\dots,x_n),\dots,f^m(x_1,\dots,x_n)),$$

then, for sake of simplicity, we put $f(x) = (f^A(x_a))$, with $1 \le a \le n$, $1 \le A \le m$.

PROPOSITION (1). Two mappings f,g from $\mathbb{R}^n$ to $\mathbb{R}^m$ have the same k-jet at x in $\mathbb{R}^n$ if and only if for every A in $(1,\dots,m)$, f^A and g^A have the same Taylor polynomial expansion at x, truncated at order k (inclusive).

Proof: Suppose that f^A and g^A have the same Taylor expansion of order k at x. Then $f(x) = g(x)$ and, if $h: \mathbb{R}^m \to \mathbb{R}$ is an arbitrary function, then the Taylor's expansion of $h\circ f$ and $h\circ g$ at x is obtained by the substitution of Taylor's expansion of f^A and g^A (at x) into the Taylor expansion of h at $f(x)$ and $g(x)$. So $h\circ f - h\circ g$ is flat or order k at x. The converse is trivial and so the proposition is proved. □

REMARK (3). It follows that all k-jets at $x \in \mathbb{R}^n$ can be identified to the m-tuples of polynomials on n variables with degree k. If $J^k(n,m)$ is the vector space of such polynomials, excluded the constant terms, then $J^k(\mathbb{R}^n,\mathbb{R}^m)$ can be identified to $\mathbb{R}^n\times\mathbb{R}^m\times J^k(n,m)$ (and, now, it is clear how one obtains the dimension of such manifold, given by (*) in §2.1, Ch. III). If $P^kf(x)$ denotes the Taylor expansion of f at x up to order k, then $\tilde{f}^k(x)\rightsquigarrow(x,f(x),P^kf(x)-f(x))$ gives the mentioned identification.

Let us see a very simple example: if $f(x) = x^3$, identifying $J^3(1,1)$ to $\mathbb{R}^3$ with the aid of $at+bt^2+ct^3 \rightsquigarrow$ $\rightsquigarrow (a,b,c)$, then the Taylor expansion of $f(x+t)$ up to order 3 is $x^3+3x^2t+3xt^2+t^3$ and so

$$\tilde{f}^3(x) = (x,f(x),P^3f(x)-f(x)) = (x,x^3,3x^2,3x,1).$$

Generally, the term of order r of the Taylor polynomial is

$$\frac{1}{r!}\left(\sum_1^m t_a \frac{\partial}{\partial x_a}\right)^r f(x) = \frac{1}{r!}\sum_1^m{}_{a_1,\ldots,a_r} t_{a_1\ldots a_r}\frac{\partial^r f}{\partial x_{a_1}\ldots\partial x_{a_r}}(x),$$

where $f\colon \mathbb{R}^n \to \mathbb{R}^m$ is smooth, $x \in \mathbb{R}^n$ and $1 \le a_1 \le\ldots\le a_r \le n$.

If we put $f(x) = (f^A(x_a)) = (y^A(x_a))$ and $y^A_{a_1\ldots a_r} =$ $= (\partial^r y^A/\partial x_{a_1}\ldots\partial x_{a_r})$, then the k-jet at x of f is represented by

$$\tilde{f}^k(x) = (x_a, y^A, y^A_{a_1\ldots a_r}). \tag{1}$$

To simplify more the notation we shall conventionate that

$$a(r) = a_1\ldots a_r .$$

For a summation index:

$$\sum_{1}^{n} a(r) = \sum_{1}^{n} a_1 \ldots a_r = \sum_{1 \le a_1 \le \ldots \le a_r \le n}$$

(in the text we shall often use the usual summation convention over repeated indices: $z_i t^i = \sum_i z_i t^i$). Also, sometimes we put $y^A = y^A_{a(0)}$. So, (1) can be written:

$$\tilde{f}^k(x) = (x_a, y^A, y^A_{a(r)}) = (x_a, y^A_{a(r)}),$$

where in the second term r runs from 1 to k and in the third from 0 to k.

2.5. Local coordinates for jets

Let N and E be manifolds of dimensions n and m, respectively. The jet manifold $J^k(N,E)$ has an atlas when the charts are modeled in the space $\mathbb{R}^n \times \mathbb{R}^m \times J^k(n,m)$ (locally, we may think $J^k(N,E)$ as $J^k(\mathbb{R}^n, \mathbb{R}^m)$). If $u(x) \in J^k(N,E)$, for $x \in N$, and $s\colon N \to E$ is such that $\tilde{s}^k(x) = u(x)$, then, for two charts (U,f) and (V,g) of N and E, respectively, with $x \in U$, $s(x) \in V$, the bijection

$$\tilde{s}^k(x) \to (f(x), g(s(x)), D^1(g \circ s \circ f^{-1})(f(x)), \ldots, D^k(g \circ s \circ f^{-1})(f(x)))$$

defines an atlas for $J^k(N,E)$, where D^j is the j-partial derivative for the corresponding Taylor expansion. So, locally, $(x_a, y^A, y^A_{a(r)})$, $1 \le r \le k$, is the coordinate representation of a point in $J^k(N,E)$: this coordinate is given for a k-jet prolongation. For the general case, we put

$$(x_a, y^A, z^A_{a(r)})$$

since we can not say that globally the sections of $(J^k E, \alpha^k, N)$

are k-prolongations of mappings from N to E. So

$$z^A_{a(r)} = y^A_{a(r)} = \frac{\partial^r y^A}{\partial x_{a_1} \cdots \partial x_{a_r}}$$

only in the case where $u(x) = \tilde{s}^k(x)$ for some $s\colon N \to E$.

In Chapter III, we shall see a way to characterize that case in terms of differential 1-forms. It is clear that the same considerations are valid for the k-jets of local maps and sections.

2.6. Holonomic, semi-holonomic and non-holonomic prolongations

Let $k \geq 1$ be an integer. Let (M,p,N) be a fibered manifold and $J^kM = J^k(M,p,N)$ the corresponding manifold of k-jets. As we have seen, J^kM is fibered over N by the source projection $\alpha^k\colon J^kM \to N$. Let $J^1(J^kM,\alpha^k,N)$ be the manifold of 1-jets of sections of $\alpha^k\colon J^kM \to N$. The major interest of such manifold is that the fibered manifold $J^{k+1}M = J^{k+1}(M,p,N)$ can be regularly immersed into $J^1(J^kM,\alpha^k,N)$ (that is, there is a differentiable mapping, injective with injective differential in every point). We do this by taking the map

$$\tilde{s}^{k+1}(x) \to \tilde{u}^1(x),$$

where $u\colon N \to J^kM$; $y \to \tilde{s}^k(y)$. In this way, every element of $J^{k+1}M$ can be considered as an element of $J^1(J^kM) = $
$= J^1(J^kM,\alpha^k,N)$.

For our future purposes, it is convenient to see this in local coordinates. For sake of simplicity, let us take

$k = 3$ and consider the coordinate systems

$$(x_a, y^A, z^A_{a(1)}, z^A_{a(2)}, z^A_{a(3)}), \quad \text{for} \quad J^3M, \tag{2}$$

$$(x_a, y^A, \bar{z}^A_{a(1)}, \bar{z}^A_{a(2)}), \quad \text{for} \quad J^2M, \tag{3}$$

$$(x_a, y^A, \bar{z}^A_{a(1)}, \bar{z}^A_{a(2)}, g^A_{,a}, g^A_{a(1),a}, g^A_{a(2),a}), \quad \text{for} \quad J^1(J^2M), \tag{4}$$

If u is in J^2M, then we have

$$\tilde{u}^1 = (x_a, y^A, \bar{z}^A_{a(1)}, \bar{z}^A_{a(2)}, \frac{\partial y^A}{\partial x_a}, \frac{\partial \bar{z}^A_{a(1)}}{\partial x_a}, \frac{\partial \bar{z}^A_{a(2)}}{\partial x_a}). \tag{5}$$

Let $s \in \mathrm{Sec}(M)$ be such that $u = \tilde{s}^2(x)$. Then

$$u = (x_a, y^A, \frac{\partial y^A}{\partial x_{a(1)}}, \frac{\partial^2 y^A}{\partial x_{a(2)}}) \in J^2M,$$

and, so, from (3)

$$\bar{z}^A_{a(1)} = \frac{\partial y^A}{\partial x_{a(1)}}, \quad \bar{z}^A_{a(2)} = \frac{\partial^2 y^A}{\partial x_{a(2)}},$$

and (5) takes the form

$$\tilde{u}^1 = (x_a, y^A, \frac{\partial y^A}{\partial x_{a(1)}}, \frac{\partial^2 y^A}{\partial x_{a(2)}}, \frac{\partial y^A}{\partial x_{a(1)}}, \frac{\partial^2 y^A}{\partial x_{a(2)}}, \frac{\partial^3 y^A}{\partial x_{a(3)}}). \tag{6}$$

But $\tilde{u}^1$ is in $J^1(J^2M)$, and so

$$g^A_{,a} = \frac{\partial y^A}{\partial x_a}, \quad g^A_{a(1),a} = \frac{\partial^2 y^A}{\partial x_{a(2)}}, \quad g^A_{a(2),a} = \frac{\partial^3 y^A}{\partial x_{a(3)}}.$$

If we take $\tilde{s}^3(x) \in J^3M$ such that

$$\tilde{s}^3(x) = (x_a, y^A, \frac{\partial y^A}{\partial x_{a(1)}}, \frac{\partial^2 y^A}{\partial x_{a(2)}}, \frac{\partial^3 y^A}{\partial x_{a(3)}}),$$

then $\tilde{s}^3(x)$ is an element of $J^1(J^2M)$ in the sense of (6):

$$\tilde{s}^3(x) = (x_a, y^A, \underbrace{\frac{\partial y^A}{\partial x_{a(1)}}, \frac{\partial^2 y^A}{\partial x_{a(2)}}}_{\text{repeated twice}}, \frac{\partial^3 y^A}{\partial x_{a(3)}})$$

(for any k, $(x_a, y^A, \underbrace{\frac{\partial y^A}{\partial x_{a(1)}}, \ldots, \frac{\partial^{k-1} y^A}{\partial x_{a(k-1)}}}_{\text{repeated twice}}, \frac{\partial^k y^A}{\partial x_{a(k)}})$).

This suggests us to define the following mapping:

$$\varphi: J^3M \to J^1(J^2M)$$

in such a way that, if $v(x) \in J^3M$, then

$$x_a(\varphi(v(x))) = x_a(v(x)),$$

$$y^A(\varphi(v(x))) = y^A(v(x)),$$

$$g^A_{,a}(\varphi(v(x))) = z^A_{a(1)}(v(x)),$$

$$g^A_{a(1),a}(\varphi(v(x))) = z^A_{a(2)}(v(x)),$$

$$g^A_{a(2),a}(\varphi(v(x))) = z^A_{a(3)}(v(x)).$$

Clearly, in the general situation for any neighborhood V of $J^1(J^{k-1}M)$ such that $V \cap \varphi(J^kM) = V^k \neq \emptyset$, we have that V^k is defined by the equations

$$\bar{z}^A_{a(1)}(\varphi(v(x))) = \ldots = \bar{z}^A_{a(k-1)}(\varphi(v(x))) = 0,$$

for all $v(x)$ such that $\varphi(v(x)) \in V^k$ and where

$$(x_a, y^A, \bar{z}^A_{a(1)}, \ldots, \bar{z}^A_{a(k-1)}, g^A_{,a}, \ldots, g^A_{a(k-1),a})$$

is a coordinate system for $J^1(J^{k-1}M)$. It is clear that φ is a regular immersion, that is, an embedding.

Let $S(J^kM)$ be the submanifold defined by the 1-jets

$\tilde{u}^1(x)$ such that

$$u(x) = j^1(\rho^k_{k-1} \circ u)(x),$$

and consider the following diagram which illustrates our situation

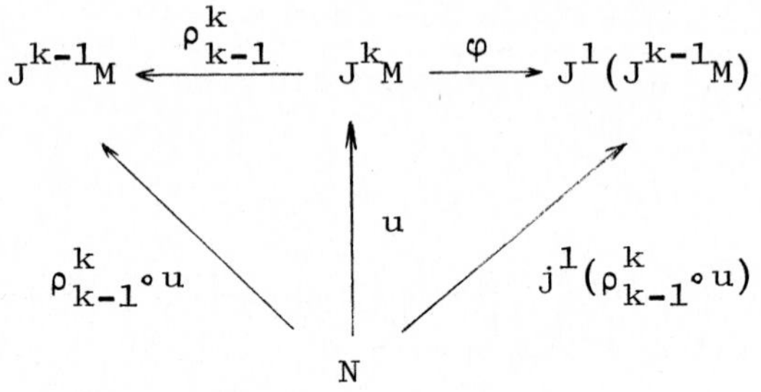

DEFINITION (5). (Ehresmann (1951)). The manifold

$J^{k+1}M$ is said to be a <u>holonomic prolongation</u> of $J^k M$

$S(J^k M)$ is said to be a <u>semi-holonomic prolongation</u> of $J^k M$

$J^1(J^k M)$ is said to be a <u>non-holonomic prolongation</u> of $J^k M$.

The following inclusions are verified:

$$J^{k+1}M \subset S(J^k M) \subset J^1(J^k M).$$

The reader can easily verify that if $(x_a, y^A, z^A_{a(r)})$ are coordinates on an open set U of $J^k M$ then $(x_a, y^A, z^A_{a(r)}, z^A_{a(r),a})$ are coordinates on $(\beta^1)^{-1}(U) \subset \subset J^1(J^k M)$, symmetrical in $a_1, \ldots, a_r$, but not necessarily in $a_1, \ldots, a_r, a$.

I.3. - Exterior differential geometry of higher order tangent bundles

3.1. The tangent bundle of higher order

The study developed in Chapter II will be made on a particular case of Jet manifolds: the tangent bundle of higher order. Let k be an arbitrary non-negative integer. Then T^kM, the tangent bundle of order k of a manifold M, in the $(k+1)m$-dimensional manifold $(m = \dim M)$ of all k-jets with source at the origin of $\mathbb{R}$ and target in M, that is,

$$T^kM = J_o^k(\mathbb{R},M) = J_o^k(\mathbb{R}\times M,p,\mathbb{R}).$$

(T^kM is a submanifold of $J^k(\mathbb{R},M)$).

One has, naturally, the same type of fibrations as was pointed out above. In fact, if $r \le k$, we have the canonical projection $\rho_r^k\colon T^kM \to T^rM$, given by $\rho_r^k(\tilde{\sigma}^k(0)) = \tilde{\sigma}^r(0)$, and the target projection $\beta^k\colon T^kM \to M$, given by $\beta^k(\tilde{\sigma}^k(0)) = \sigma(0)$. Obviously, one has $\rho_o^k = \beta^k$ (where T^oM is canonically identified to M).

Let us remark that if $k = 1$, then we are in presence of the usual tangent bundle of the manifold M. In this case, the projection β^1 will be denoted by p_M.

We shall now describe the local coordinates in T^kM. Let U be a chart of M with local coordinates y^A, $1\le A\le m$, $\sigma\colon\mathbb{R} \to M$ a curve in M such that $\sigma(0) \in U$ and put $\sigma^A = y^A\circ\sigma$, $1\le A\le m$. Then the k-jet $\tilde{\sigma}^k(0)$ is uniquely represented in $(\beta^k)^{-1}(U) = T^kU$ by

$$(y^A, z^A_{(1)}, \dots, z^A_{(k)}), \qquad 1 \le A \le m, \tag{1}$$

where

$$y^A = \sigma^A(0), \quad z^A_{(i)} = \frac{1}{i!}\frac{d^i\sigma^A}{dt^i}(0), \qquad 1 \le i \le k.$$

Sometimes we shall employ the notation $z^A_{(0)} = y^A$. Then we have a chart $(\beta^k)^{-1}(U)$ in T^kM with local coordinates $(z^A_{(0)}, z^A_{(1)}, \ldots, z^A_{(k)})$. The factor $\frac{1}{i!}$ appears only for a technical reason, because the differential calculus developed in this chapter becomes simpler. However, from a physical viewpoint, the choice of these local coordinates is not the more adequate (for instance, the acceleration of a curve σ is given by $\frac{d^2\sigma^A}{dt^2}$). Then we shall sometimes consider the following coordinate system in $(\beta^k)^{-1}(U)$:

$$(q^A, q^{(1)A}, \ldots, q^{(k)A}), \qquad 1 \le A \le m,$$

where $q^A = \sigma^A(0)$, $q^{(i)A} = \frac{d^i\sigma^A}{dt^i}(0)$, $1 \le i \le k$.

We put $q^{(0)A} = q^A$, $\dot{q}^A = q^{(1)A}$, $\ddot{q}^A = q^{(2)A}$, and so on.

Then we have $q^{(i)A} = i!\ z^A_{(i)}$, $0 \le i \le k$, $1 \le A \le m$.

Now, let σ be a curve in M. We shall denote by $\tilde{\sigma}^k$ the <u>canonical prolongation</u> of σ to T^kM defined as follows:

$$\tilde{\sigma}^k(t) = \bar{\sigma}^k_t(0),$$

where $\sigma_t(s) = \sigma(s+t)$. If $k = 1$, we put $\tilde{\sigma}^1 = \dot{\sigma}$. Along the prolongation $\tilde{\sigma}^k$ to T^kM of a curve σ in M we have

$$q^{(i)A} = \frac{d^i\sigma^A}{dt^i}, \qquad 0 \le i \le k.$$

We shall employ the coordinates $(q^{(0)A}, \ldots, q^{(k)A})$ in Chapter II when jet prolongation of curves are involved.

Now, let $z \in T^kM$ and $X \in T_z(T^kM)$ be a tangent vector with z locally given by (1). Then X is given by

$$X = (y^A, z^A_{(i)}, X_A, X^{(i)}_A), \quad 1 \le A \le m, \quad 1 \le i \le k,$$

or

$$X = X_A \frac{\partial}{\partial y^A} + X^{(1)}_A \frac{\partial}{\partial z^A_{(1)}} + \ldots + X^{(k)}_A \frac{\partial}{\partial z^A_{(k)}}$$

(where the summation convention is adopted).

If $r \le k$, the projection $\rho^k_r : T^kM \to T^rM$ is locally given by

$$(z^A_{(0)}, z^A_{(1)}, \ldots, z^A_{(k)}) \rightsquigarrow (z^A_{(0)}, z^A_{(1)}, \ldots, z^A_{(r)}), \tag{2}$$

and the prolongated projection (or tangent map) $T\rho^k_r$ is locally given by

$$(z^A_{(0)}, z^A_{(1)}, \ldots, z^A_{(k)}, X^{(0)}_A, X^{(1)}_A, \ldots, X^{(k)}_A) \to$$
$$\rightsquigarrow (z^A_{(0)}, z^A_{(1)}, \ldots, z^A_{(r)}, X^{(0)}_A, X^{(1)}_A, \ldots, X^{(r)}_A). \tag{3}$$

Now, we shall see how each fibered manifold $\rho^k_{r-1} : T^kM \to T^{r-1}M$, $1 \le r \le k$, determines an exact sequence of vector bundles over T^kM. In order to do this, we first consider the vertical bundle of T^kM over $T^{r-1}M$ which will be denoted by $V^{\rho^k_{r-1}}(T^kM)$. Then $V^{\rho^k_{r-1}}(T^kM)$ consists of all tangent vectors to T^kM which are projected onto 0 by $T\rho^k_{r-1}$. Taking into account (2) and (3), if z is a point in T^kM and X an element of $V^{\rho^k_{r-1}}(T^kM)$ at z, then X is locally given by

$$X = (z^A_{(0)}, \ldots, z^A_{(k)}, 0, \ldots, 0, X^{(r)}_A, \ldots, X^{(k)}_A). \tag{4}$$

If we denote $i_{k-r+1} : V^{\rho^k_{r-1}}(T^kM) \to T(T^kM)$ the canonical inclusion, then i_{k-r+1} is locally given by

$$(a^A_{(0)},\ldots,z^A_{(k)},X^{(r)}_A,\ldots,X^{(k)}_A) \to$$
$$\rightsquigarrow (z^A_{(0)},\ldots,z^A_{(k)},0,\ldots,0,X^{(r)}_A,\ldots,X^{(k)}_A). \tag{5}$$

Let us consider the induced bundle of $p_{T^{r-1}M} : T(T^{r-1}M) \to T^{r-1}M$ via ρ^k_{r-1}, denoted by $T^kM \times_{T^{r-1}M} T(T^{r-1}M)$; recall that $T^kM \times_{T^{r-1}M} T(T^{r-1}M)$ is the set of ordered pairs (z,X) of $T^kM \times T(T^{r-1}M)$ such that $\rho^k_{r-1}(z) = p_{T^{r-1}M}(X)$. Then $T^kM \times_{T^{r-1}M} T(T^{r-1}M)$ is a vector bundle over T^kM and the following diagram

$$\begin{array}{ccc} T^kM \times_{T^{r-1}M} T(T^{r-1}M) & \dashrightarrow & T(T^{r-1}M) \\ \Big\downarrow & & \Big\downarrow p_{T^{r-1}M} \\ T^kM & \xrightarrow{\ \rho^k_{r-1}\ } & T^{r-1}M \end{array}$$

is commutative, where the dotted arrows are the canonical projections (for more details of the construction above, we remit the reader to the Appendix A; also, the book of Godbillon (1969) can be useful).

Moreover, we have the following commutative diagram:

$$\begin{array}{ccc} T(T^kM) & \xrightarrow{\ T\rho^k_{r-1}\ } & T(T^{r-1}M) \\ \Big\downarrow p_{T^kM} & & \Big\downarrow p_{T^{r-1}M} \\ T^kM & \xrightarrow{\ \rho^k_{r-1}\ } & T^{r-1}M \end{array}$$

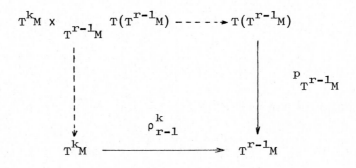

Therefore, we deduce

PROPOSITION (1). <u>There is only one vector bundle homomorphism $s_{k-r+1}: T(T^kM) \to T^kM \times_{T^{r-1}M} T(T^{r-1}M)$ over T^kM such that the following diagram:</u>

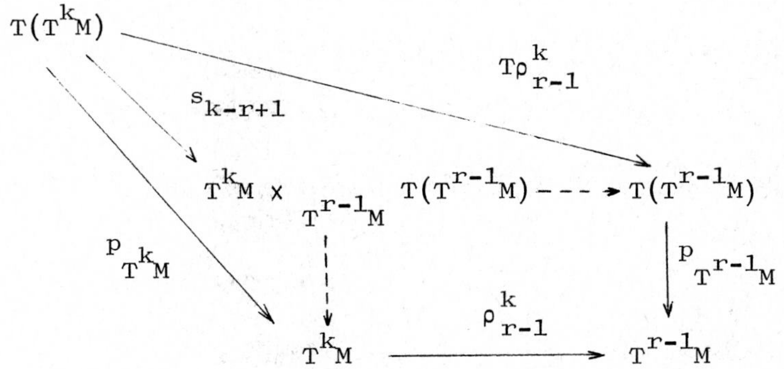

<u>is commutative.</u>

Proof: Define s_{k-r+1} by

$$s_{k-r+1}(X) = (p_{T^kM}(X), T\rho^k_{r-1}(X)),$$

for every tangent vector X to T^kM. Then s_{k-r+1} is locally given by

$$(z^A_{(i)}, X^{(i)}_A) \rightsquigarrow (z^A_{(0)},\ldots,z^A_{(k)}; z^A_{(0)},\ldots,z^A_{(r-1)}, X^{(0)}_A,\ldots,X^{(r-1)}_A). \quad (6)$$

From this local expression, one easily deduces that such s_{k-r+1} exists and is unique. □

Using (5) and (6), it is not hard to show that s_{k-r+1} is onto and that $\operatorname{Ker} s_{k-r+1} = V^{\rho^k_{r-1}}(T^kM)$. Consequently, we obtain the following exact sequence of vector bundles over T^kM:

$$0 \to V^{\rho^k_{r-1}}(T^kM) \xrightarrow{i_{k-r+1}} T(T^kM) \xrightarrow{s_{k-r+1}} T^kM \times_{T^{r-1}M} T(T^{r-1}M) \to 0 \quad (7)$$

which is locally given by

$$0 \to (z^A_{(0)},\ldots,z^A_{(k)},X^{(r)}_A,\ldots,X^{(k)}_A) \leadsto (z^A_{(0)},\ldots,z^A_{(k)},0,\ldots,0,X^{(r)}_A,\ldots,X^{(k)}_A)$$

$$(z^A_{(0)},\ldots,z^A_{(k)},X^{(0)}_A,\ldots,X^{(k)}_A) \leadsto (z^A_{(0)},\ldots,z^A_{(k)};z^A_{(0)},\ldots,z^A_{(r-1)},X^{(0)}_A,\ldots,X^{(r-1)}_A) \to 0$$

DEFINITION (1). The exact sequence of vector bundles (7) will be called the (k-r+1)-th <u>fundamental exact sequence</u>.

Summarizing, we have constructed k exact sequences of vector bundles over T^kM:

1st $\quad 0 \to V^{\rho^k_{k-1}}(T^kM) \xrightarrow{i_1} T(T^kM) \xrightarrow{s_1} T^kM \times_{T^{k-1}M} T(T^{k-1}M) \to 0$

. . . .

rth $\quad 0 \to V^{\rho^k_{k-r}}(T^kM) \xrightarrow{i_r} T(T^kM) \xrightarrow{s_r} T^kM \times_{T^{k-r}M} T(T^{k-r}M) \to 0$

. . . .

kth $\quad 0 \to V^{\beta^k}(T^kM) \xrightarrow{i_k} T(T^kM) \xrightarrow{s_k} T^kM \times_M TM \to 0$

These k fundamental exact sequences can be connected as follows. Let us define a mapping

$$h_r : T^kM \times_{T^{r-1}M} T(T^{r-1}M) \to V^{\rho^k_{k-r}}(T^kM)$$

by

$$(z^A_{(0)},\ldots,z^A_{(k)};z^A_{(0)},\ldots,z^A_{(r-1)},X^{(0)}_A,\ldots,X^{(r-1)}_A) \leadsto$$
$$\leadsto (z^A_{(0)},\ldots,z^A_{(k)},0,\ldots,0,X^{(0)}_A,\ldots,X^{(r-1)}_A). \tag{8}$$

We can easily check that h_r is globally well defined and, in fact, a vector bundle isomorphism over T^kM. Then, using these isomorphisms, h_r, $1 \le r \le k$, the r th and the (k-r+1) th fundamental exact sequences are connected as follows:

$$
\begin{array}{llllllll}
\text{rth} & 0 \to V^{\rho^k_{k-r}}(T^kM) & \xrightarrow{i_r} & T(T^kM) & \xrightarrow{s_r} & T^kM\times_{T^{k-r}M} T(T^{k-r}M) & \to 0 \\
 & \quad\nwarrow^{h_r} & & & & \swarrow_{h_{k-r+1}} & \\
(k-r+1)\text{th} & 0 \to V^{\rho^k_{r-1}}(T^kM) & \xrightarrow{i_{k-r+1}} & T(T^kM) & \xrightarrow{s_{k-r+1}} & T^kM\times_{T^{r-1}M} T(T^{r-1}M) & \to 0
\end{array}
$$

Next, in order to make this construction above more understandable, we shall particularize it for $k = 1$ and $k = 2$.

Let us take $k = 1$. Then there exists one exact sequence of vector bundles over TM

$$0 \to V(TM) \xrightarrow{i} T(TM) \xrightarrow{s} TM \times_M TM \to 0 \tag{9}$$

locally given by

$$0 \to (z^A_{(0)}, z^A_{(1)}, x^{(1)}_A) \rightsquigarrow (z^A_{(0)}, z^A_{(1)}, 0, x^{(1)}_A)$$

$$(z^A_{(0)}, z^A_{(1)}, x^{(0)}_A, x^{(1)}_A) \rightsquigarrow (z^A_{(0)}, z^A_{(1)}; z^A_{(0)}, x^{(0)}_A) \to 0$$

and one vector bundle isomorphism over TM

$$h\colon TM \times_M TM \to V(TM)$$

locally given by

$$(z^A_{(0)}, z^A_{(1)}; z^A_{(0)}, x^{(0)}_A) \rightsquigarrow (z^A_{(0)}, z^A_{(1)}, 0, x^{(0)}_A). \tag{10}$$

(See Vilms (1967), Klein & Voutier (1968), Godbillon (1969)).

Let us now take $k = 2$. Then T^2M is fibered over M and TM, respectively:

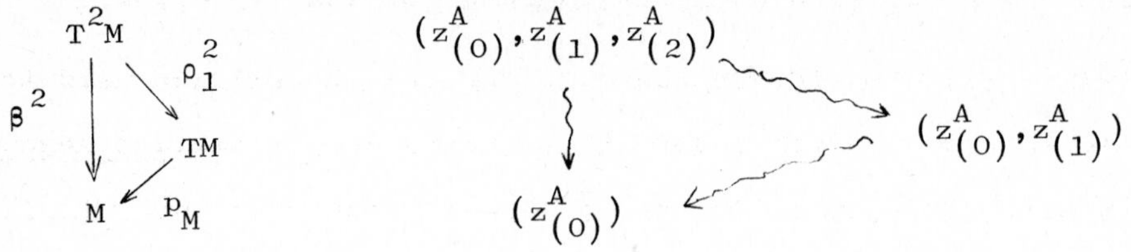

Each fibration aboves gives an exact sequence of vector bundles over T^2M:

$$\left.\begin{array}{ll} \text{1st} & 0 \to V^{\rho^2_1}(T^2M) \xrightarrow{i_1} T(T^2M) \xrightarrow{s_1} T^2M \times_{TM} TTM \to 0 \\ \text{2nd} & 0 \to V^{\beta^2}(T^2M) \xrightarrow{i_2} T(T^2M) \xrightarrow{s_2} T^2M \times_M TM \to 0 \end{array}\right\} \quad (11)$$

locally given by

$$0 \to (z^A_{(0)}, z^A_{(1)}, z^A_{(2)}, X^{(2)}_A) \rightsquigarrow (z^A_{(0)}, z^A_{(1)}, z^A_{(2)}, 0, 0, X^{(2)}_A)$$

$$\text{1st} \quad (z^A_{(0)}, z^A_{(1)}, z^A_{(2)}, X^{(0)}_A, X^{(1)}_A, X^{(2)}_A) \to (z^A_{(0)}, z^A_{(1)}, z^A_{(2)}, z^A_{(0)}, z^A_{(1)}, X^{(0)}_A, X^{(1)}_A) \to 0$$

$$0 \to (z^A_{(0)}, z^A_{(1)}, z^A_{(2)}, X^{(1)}_A, X^{(2)}_A) \rightsquigarrow (z^A_{(0)}, z^A_{(1)}, z^A_{(2)}, 0, X^{(1)}_A, X^{(2)}_A)$$

$$\text{2nd} \quad (z^A_{(0)}, z^A_{(1)}, z^A_{(2)}, X^{(0)}_A, X^{(1)}_A, X^{(2)}_A) \to (z^A_{(0)}, z^A_{(1)}, z^A_{(2)}, z^A_{(0)}, X^{(0)}_A) \to 0$$

Hence, there exist two vector bundle isomorphisms over T^2M

$$h_1 : T^2M \times_M TM \to V^{\rho^2_1}(T^2M)$$

and

$$h_2 : T^2M \times_{TM} TTM \to V^{\beta^2}(T^2M)$$

locally given by

$$(z^A_{(0)}, z^A_{(1)}, z^A_{(2)}; z^A_{(0)}, X^{(0)}_A) \rightsquigarrow (z^A_{(0)}, z^A_{(1)}, z^A_{(2)}, 0, 0, X^{(0)}_A) \qquad (12)_1$$

and

$$(z^A_{(0)}, z^A_{(1)}, z^A_{(2)}; z^A_{(0)}, z^A_{(1)}, X^{(0)}_A, X^{(1)}_A) \rightsquigarrow (z^A_{(0)}, z^A_{(1)}, z^A_{(2)}, 0, X^{(0)}_A, X^{(1)}_A) \qquad (12)_2$$

respectively. (See Catz (1974)(a) and de León (1978), (1981)).

REMARK (1). The tangent bundle of order k will be used to describe the Generalized Particle Mechanics in the <u>autonomous</u> sense (that is, the time appears in the functions only implicity). For a non-autonomous Mechanics we consider the manifold $J^k(\mathbb{R},M)$ of jets of order k with local coordinates $(t,y^A,z^A_{(i)})$ (or, if we want, $\mathbb{R}\times T^kM$).

REMARK (2). We can consider a more general framework in order to develop a Classical Field Theory in the autonomous sense. Indeed, let T^k_pM be the manifold of all k-jets with the origin of $\mathbb{R}^p$ as source and target in M, that is,

$$T^k_pM = J^k_o(\mathbb{R}^p,M) = J^k_o(\mathbb{R}^p\times M,p,\mathbb{R}^p).$$

The manifold T^k_pM was introduced by Ehresmann and will be called the <u>tangent bundle of</u> p^k<u>-velocities</u> (see Ehresmann (1955)(b), Morimoto (1969), (1970), Tulczyjew (1977)). We have local coordinates $(y^A,z^A_{a(r)})$ in T^k_pM, $1 \leq A \leq \dim M$, $1 \leq r \leq k$. Obviously, for $p = 1$, we are in presence of the tangent bundle of order k of M. The non-autonomous case could be described by using the manifold $J^k(\mathbb{R}^p,M)$, or, if we want, $\mathbb{R}^p \times T^k_pM$, with local coordinates $(t_a,z^A,z^A_{a(r)})$. Notice that all constructions above can be made for T^k_pM without major difficulties.

3.2. <u>The Liouville vector field of higher order</u>

The specific result presented in this section is the definition of a vector field on T^kM which generalizes the Liouville vector field canonically defined on the tangent bundle TM (See Godbillon (1969)). More precisely, we shall see that each fundamental exact sequence yields a canonical vector field on T^kM. We proceed as follows. First, notice that there exists a canonical mapping $j_r : T^kM \to T(T^{r-1}M)$, $1 \leq r \leq k$, defined by

$$\tilde{\sigma}^k(0) \rightsquigarrow \tilde{\tau}^1(0),$$

where $\tau : \mathbb{R} \to T^{r-1}M$, $t \to \tau(t) = \tilde{\sigma}_t^{r-1}(0)$, and $\sigma_t(s) = \sigma(s+t)$. It is convenient to describe j_r in local coordinates. A simple computation show that j_r is locally given by

$$(z^A_{(0)},\ldots,z^A_{(k)}) \to (z^A_{(0)},\ldots,z^A_{(r-1)},z^A_{(1)},2z^A_{(2)},\ldots,rz^A_{(r)}). \qquad (13)$$

Let us now consider the vector field C_r on T^kM defined by the following composition:

$$T^kM \xrightarrow{\;Id \times j_{k-r+1}\;} T^kM \times_{T^{k-r}M} T(T^{k-r}M) \xrightarrow{\;h_{k-r+1}\;} V^{\rho^k_{r-1}}(T^kM),$$

$$C_r : T^kM \to V^{\rho^k_{r-1}}(T^kM)$$

that is,

$$C_r = h_{k-r+1} \circ (Id \times j_{k-r+1}).$$

From (8) and (13), we deduce that C_r is locally given by

$$C_r(z^A_{(0)},\ldots,z^A_{(k)}) = (z^A_{(0)},\ldots,z^A_{(k)};0,\ldots,0,z^A_{(1)},2z^A_{(2)},\ldots,(k-r+1)\,z^A_{(k-r+1)}),$$

or, else,

$$\begin{aligned} C_r &= \sum_{i=1}^{k-r+1} i z^A_{(i)} \frac{\partial}{\partial z^A_{(r+i-1)}} \\ &= z^A_{(1)} \frac{\partial}{\partial z^A_{(r)}} + 2z^A_{(2)} \frac{\partial}{\partial z^A_{(r+1)}} + \ldots + (k-r+1) z^A_{(k-r+1)} \frac{\partial}{\partial z^A_{(k)}} \end{aligned} \qquad (14)$$

Notice that, for the case $k = 1$, $j_1 : TM \to TM$ is nothing but the identity map; then, there is one vector field on TM which is locally given by

$$C = z^A_{(1)} \frac{\partial}{\partial z^A_{(1)}}$$

or, introducing the usual notation $(q^A, \dot{q}^A)$ for the local coordinates in TM,

$$C = \dot{q}^A \frac{\partial}{\partial \dot{q}^A}. \tag{15}$$

So, we are in presence of the Liouville vector field on TM.

For the case $k = 2$, we obtain two canonical vector fields on T^2M, C_1 and C_2, locally given by

$$C_1 = z^A_{(1)} \frac{\partial}{\partial z^A_{(1)}} + 2z^A_{(2)} \frac{\partial}{\partial z^A_{(2)}} \qquad (16)_1$$

$$C_2 = z^A_{(1)} \frac{\partial}{\partial z^A_{(2)}} \qquad (16)_2$$

(see Catz (1974) and de León (1978), (1981)).

DEFINITION (2). The vector field C_r will be called the r th <u>canonical vector field</u> on T^kM and C_1 the <u>Liouville vector field of order</u> k.

Here is a key geometrical property of the Liouville vector field of order k: its flow consists of the homothetias of T^kM. Indeed, let $h_t : \mathbb{R} \to \mathbb{R}$ be the homothetia of ratio e^t, that is, $h_t(s) = e^t s$, and let $H_t : T^kM \to T^kM$ denote the fiber preserving transformation deduced from h_t, that is

$$H_t(\tilde{\sigma}^k(0)) = (\widetilde{\sigma \circ h_t})^k(0).$$

Therefore, H_t is locally given by

$$(z^A_{(0)}, z^A_{(1)}, \ldots, z^A_{(k)}) \rightsquigarrow (z^A_{(0)}, e^t z^A_{(1)}, \ldots, e^{kt} z^A_{(k)}). \tag{17}$$

From (14) and (17), one easily proves the following

PROPOSITION (2). H_t _is the flow (or the one-parameter group) generated by_ C_1.

REMARK (3). This geometrical property is well known for the usual Liouville vector field on TM (see Godbillon (1969), for $k = 1$ and Catz (1974) and de León (1978), (1981), for $k = 2$).

3.3. The almost tangent structure of higher order

Our aim in this paragraph is to study a certain geometric structure canonically associated to the tangent bundle of order k of a differentiable manifold which generalizes, in an obvious manner, the so-called almost tangent structure on the tangent bundle of a manifold.

First, let us recall some well known facts about this kind of geometric structures.

The notion of an almost tangent structure on a 2m-dimensional manifold appears as a natural generalization of the canonical structure determined on the tangent bundle TM of a manifold M. The study of these structures was initiated by Clark & Bruckheimer (1960) and further developed by many authors, for example, Eliopoulos (1962), (1965), Houh (1969), Clark & Goel (1972) and Yano & Davies (1975).

DEFINITION (3). Let E be a 2m-dimensional manifold. If E

admits an endomorphism J of TE such that

$$J^2 = 0 \quad \text{and} \quad \text{rank } J = m \quad (\text{all over } E),$$

then J is called an <u>almost tangent structure on</u> E.

Let us now consider the tangent bundle TM of a manifold M and the endomorphism J of $T(TM)$ given by $J = i\circ h\circ s$. From (9) and (10), one easily proves that J is locally given by

$$(z^A_{(0)}, z^A_{(1)}, X^{(0)}_A, X^{(1)}_A) \rightarrow (z^A_{(0)}, z^A_{(1)}, 0, X^{(0)}_A), \tag{18}$$

and so, the matrix representation of J is

$$\begin{pmatrix} 0 & 0 \\ I_m & 0 \end{pmatrix},$$

where I_m denotes the unit matrix of order m, $\dim M = m$. Taking into account (18), we deduce that J defines an almost tangent structure on TM, which will be called the <u>canonical almost tangent structure</u> on TM (for more details, see Godbillon (1969)).

These kind of geometric structures on manifolds was generalized by Eliopoulos (1966) taking as model the tangent bundle of higher order of a manifold.

DEFINITION (4). Let E be a $(k+1)m$-dimensional manifold. If E admits an endomorphism J of TE such that

$$J^{k+1} = 0 \quad \text{and} \quad \text{rank } J = km \quad (\text{all over } E),$$

then J is called an <u>almost tangent structure of order</u> k <u>on</u> E.

REMARK (4). The almost tangent structures of order 2 were extensively studied by Clark & Goel (1972) and Catz (1974). Notice that the almost tangent structures of higher order can be considered as a particular case of certain geometric structures early introduced by Lehmann-Lejeune (1964).

As we point out above, the notion of an almost tangent structure of order k is the natural generalization of the canonical structure determined on the tangent bundle of order k, T^kM, of a manifold M, which will be described next.

DEFINITION (5). The endomorphism

$$J_r = i_{k-r+1} \circ h_{k-r+1} \circ s_r , \quad 1 \leq r \leq k,$$

of $T(T^kM)$ will be called the r th <u>vertical endomorphism</u> of $T(T^kM)$.

Then we have k endomorphisms of $T(T^kM)$, $J_1,\ldots,J_k$.

The following proposition gives the relation between them.

PROPOSITION (3). <u>The</u> r-th <u>vertical endomorphism</u> J_r <u>has constant rank</u> rm <u>and satisfies</u>

$$(J_r)^s = \begin{cases} 0 , & \text{if } rs \geq k+1 \\ J_{rs} , & \text{if } rs < k+1 \end{cases}$$

<u>Proof</u>: From (5), (6) and (8), we deduce that J_r is locally given by

$$(z^A_{(i)}, X^{(i)}_A) \rightarrow (z^A_{(0)},\ldots,z^A_{(k)},0,\ldots,0,X^{(0)}_A,\ldots,X^{(k-r)}_A) \qquad (19)$$

and so the matrix representation of J_r is

$$\begin{pmatrix} 0 & 0 \\ I_{(k-r+1)m} & 0 \end{pmatrix}$$

where $I_{(k-r+1)m}$ denotes the unit matrix of order $(k-r+1)m$, $\dim M = m$. Now, the result follows from a direct calculation. □

REMARK (5). According to Proposition (3), the 1 st vertical endomorphism J_1 of $T(T^kM)$ has constant rank km and satisfies $(J_1)^{k+1} = 0$; then J_1 determines on T^kM an almost tangent structure of order k which will be called the <u>canonical almost tangent structure of order</u> k on T^kM. Let us take $k = 2$ for a moment. Then we have two endomorphisms of $T(T^2M)$, J_1 and J_2. From (11), $(12)_1$ and $(12)_2$, we deduce that J_1 and J_2 are locally given by

$$J_1 : (z^A_{(0)}, z^A_{(1)}, z^A_{(2)}, X^{(0)}_A, X^{(1)}_A, X^{(2)}_A) \rightsquigarrow (z^A_{(0)}, z^A_{(1)}, z^A_{(2)}, 0, X^{(0)}_A, X^{(1)}_A) \quad (20)_1$$

$$J_2 : (z^A_{(0)}, z^A_{(1)}, z^A_{(2)}, X^{(0)}_A, X^{(1)}_A, X^{(2)}_A) \rightsquigarrow (z^A_{(0)}, z^A_{(1)}, z^A_{(2)}, 0, 0, X^{(0)}_A) \quad (20)_2$$

respectively, and, hence, they have the matrix representation

$$J_1 = \begin{pmatrix} 0 & 0 & 0 \\ I_m & 0 & 0 \\ 0 & I_m & 0 \end{pmatrix}, \qquad J_2 = \begin{pmatrix} 0 & 0 & 0 \\ 0 & 0 & 0 \\ I_m & 0 & 0 \end{pmatrix},$$

respectively. So, one has

$$\operatorname{rank} J_1 = 2m \quad \text{and} \quad J_1^3 = 0,$$

and, consequently, J_1 determines an almost tangent structure of order 2 on T^2M.

REMARK (6). There is an alternative construction of the vertical endomorphisms of $T(T^kM)$ by using the theory of lifts of tensor fields on M to T^kM. We shall briefly describe this procedure (for more details, see Yano & Ishihara (1973), Morimoto (1969) and Youen (1977)). We will first define the lifts of a function f on M to T^kM. For each integer r, $0 \le r \le k$, we define the r-<u>lift</u> $f^{(r)}$ of a function f on M to T^kM by

$$f^{(r)}(\tilde{\sigma}^k(0)) = \frac{1}{r!}\frac{d^r(f\circ\sigma)}{dt^r}(0).$$

Let us now consider a vector field X on M. Then the r-<u>lift</u> of X to T^kM, $X^{(r)}$, is the unique vector field on T^kM such that

$$X^{(r)}f^{(s)} = (Xf)^{(r+s-k)}, \qquad 0 \le s \le k,$$

for any function f on M and where $f^{(s)} = 0$, if s is a negative integer. It is not hard to see that $X^{(r)}$ is locally given by

$$X^{(r)} = \sum_{s=0}^{r} (X^A)^{(s)} \frac{\partial}{\partial z^A_{(r-s)}},$$

where $X = X^A \dfrac{\partial}{\partial y^A}$ is a chart U on M.

We are now able to define the lifts of a tensor field F of type (1,1) on M (that is, an endomorphism of TM) to T^kM; the r-<u>lift</u> $F^{(r)}$ of F to T^kM is the unique tensor field of type (1,1) on T^kM (that is, the endomorphism of $T(T^kM)$) such that

$$F^{(r)}X^{(s)} = (FX)^{(r+s-k)}, \qquad 0 \le s \le k,$$

for any vector field X on M. As above, we put $X^{(s)} = 0$, if s is a negative integer.

We leave to the reader to check that, if Id is the

identity endomorphism of TM, then

$$Id^{(r)} = J_r, \quad \text{for} \quad 1 \leq r \leq k,$$

and

$$Id^{(0)} = \text{Identity endomorphism of } T(T^kM).$$

We indicate to the reader a different approach for the construction of the generalized Liouville vector field and the almost tangent structure of higher order (see Rodrigues (1975)).

From (4) and (19), it is clear that

$$\text{Ker } J_r = V^{\rho^k_{k-r}}(T^kM) = \text{Im } J_{k-r+1} \tag{21}$$

and

$$\text{Im } J_r \supset \text{Im } J_s, \quad r \leq s. \tag{22}$$

Consequently, we have a decreasing sequence of sub-bundles of $T(T^kM)$

$$\text{Im } J_1 \supset \text{Im } J_2 \supset \ldots \supset \text{Im } J_k.$$

Next, we relate the canonical vector fields to the vertical endomorphisms of T^kM. Take $k = 1$; then, a simple calculation from (15) and (18), shows that $JC = 0$. If we take $k = 2$, we have $J_1C_1 = C_2$, $J_1C_2 = J_2C_1 = J_2C_2 = 0$, because $(16)_1$, $(16)_2$, $(20)_1$ and $(20)_2$. Let us now turn to the general case. We have

PROPOSITION (4).

(i) $$J_rC_s = \begin{cases} 0, & \text{if } r+s \geq k+1 \\ C_{r+s}, & \text{if } r+s < k+1 \end{cases}$$

(ii) $$[C_r, J_s] = \begin{cases} 0, & \text{if } r+s > k+1 \\ -s\, J_{r+s-1}, & \text{if } r+s \leq k+1 \end{cases}$$

(iii) $[J_r, J_s] = 0,$

where $1 \leq r \cdot s \leq k$.

<u>Proof</u>: (i) is a direct consequence of (14) and (19). In order to prove (ii), it suffices to recall that $[C_r, J_s]$ is defined by

$$[C_r, J_s](X) = [C_r, J_s X] - J_s[C_r, X],$$

for any vector field X on T^kM. Now, (ii) follows again from (14) and (19). Finally, let us recall that $[J_r, J_s]$ is defined by

$$[J_r, J_s](X,Y) = [J_r X, J_s Y] + [J_s X, J_r Y] + J_r J_s[X,Y] + \\ + J_s J_r[X,Y] - J_r[X, J_s Y] - J_r[J_s X, Y] - J_s[X, J_r Y] - J_s[J_r X, Y],$$

for any vector fields X, Y on T^kM. Again, (iii) is a direct consequence of the local expressions of J_r and J_s. □

Let J_r^* be the <u>adjoint</u> of J_r, $1 \leq r \leq k$, that is, J_r^* is the endomorphism of the exterior algebra $\Lambda(T^kM)$ of T^kM given by

$$J_r^* \omega(X_1, \ldots, X_p) = \omega(J_r X_1, \ldots, J_r X_p),$$

where ω is a p-form and $X_1, \ldots, X_p$ are vector fields on T^kM. We put $J_r^* f = f$, for any function f on T^kM. J_r^* will be called the r-th <u>vertical operator</u> on T^kM and it is locally given by

$$\left.\begin{array}{l} J_r^* f = f, \quad \text{for any function } f \text{ on } T^kM, \\ J_r^*(dz^A_{(i)}) = \begin{cases} 0 & , \text{ if } i < r \\ dz^A_{(i-r)} & , \text{ if } i \geq r \end{cases} \end{array}\right\} \qquad (23)$$

PROPOSITION (5). We have

$$(J_r^*)^s = \begin{cases} 0 , & \text{if} \quad rs \geq k+1 \\ J_{rs}^* , & \text{if} \quad rs < k+1 \end{cases}$$

Proof: Directly from Proposition 3. □

PROPOSITION (6). We have

$$i_X \circ J_r^* = J_r^* \circ i_{J_r X} ,$$

for any vector field X on $T^k M$, $1 \leq r \leq k$.

Proof: Let $X_1, \ldots, X_{p-1}$ be vector fields and ω a p-form on $T^k M$. Then we have

$$((i_X \circ J_r^*)\omega)(X_1, \ldots, X_{p-1}) = (J_r^*\omega)(X, X_1, \ldots, X_{p-1})$$

$$= \omega(J_r X, J_r X_1, \ldots, J_r X_{p-1})$$

$$= (i_{J_r X}\omega)(J_r X_1, \ldots, J_r X_{p-1})$$

$$= ((J_r^* \circ i_{J_r X})\omega)(X_1, \ldots, X_{p-1}). \quad \square$$

Taking into account Propositions (4) and (6), one easily deduces

COROLLARY (1). We have

$$i_{C_s} \circ J_r^* = \begin{cases} 0 , & \text{if} \quad r+s \geq k+1 \\ J_r^* \circ i_{C_{r+s}} , & \text{if} \quad r+s < k+1 \end{cases}$$

Let us take $k = 1$. Then, from Corollary 1, we obtain $i_C \circ J^* = 0$. For $k = 2$, we have $i_{C_2} \circ J_2^* = i_{C_2} \circ J_1^* = i_{C_1} \circ J_2^* = 0$ and $i_{C_1} \circ J_1^* = J_1^* \circ i_{C_2}$.

3.4. Vertical derivations

This paragraph and the following ones are devoted to develop the differential calculus on higher tangent bundles determined by the corresponding almost tangent structures of higher order.

Let ω be a p-form on T^kM, $p \geq 1$. Then we can define the inner product of J_r by ω as the p-form on T^kM defined by

$$i_{J_r}\omega(X_1,\ldots,X_p) = \sum_{i=1}^{p} \omega(X_1,\ldots,J_rX_i,\ldots,X_p).$$

We put $i_{J_r} f = 0$, for any function f on T^kM.

We next recall some well known facts about derivations on manifolds. For more details see, for example, the books of Godbillon (1969) and Kobayashi & Nomizu (1963).

For our purposes, let $\Lambda(M) = \bigoplus_{p\geq 0} \Lambda^p(M)$ be the exterior algebra of an arbitrary manifold M, where $\Lambda^p(M)$ is the real vector space of p-forms on M. A derivation (resp., skew-derivation) on $\Lambda(M)$ is a linear mapping D of $\Lambda(M)$ into itself which satisfies

$$D(\omega\wedge\omega') = D\omega \wedge \omega' + \omega \wedge D\omega', \quad \text{for } \omega,\omega' \in \Lambda(M)$$

(resp., $D(\omega\wedge\omega') = D\omega \wedge \omega' + (-1)^p\omega \wedge D\omega'$, for $\omega \in \Lambda^p(M)$, $\omega' \in \Lambda(M)$).

A derivation or a skew-derivation D on $\Lambda(M)$ is said to be of degree ℓ if it maps $\Lambda^p(M)$ into $\Lambda^{p+\ell}(M)$, for every p. The exterior differential d is a skew-derivation of degree 1. A general result on derivations and skew-derivations on $\Lambda(M)$ is

PROPOSITION (7) (Kobayashi and Nomizu (1963)).

(i) If D and D' are derivations of degree ℓ and ℓ', respectively, then $[D,D'] = DD'-D'D$ is a derivation of degree $\ell + \ell'$.

(ii) If D is a derivation of degree ℓ and D' is a skew-derivation of degree ℓ', then $[D,D'] = DD'-D'D$ is a skew-derivation of degree $\ell + \ell'$.

(iii) If D and D' are skew-derivations of degree ℓ and ℓ', respectively, then $[D,D'] = DD'+D'D$ is a derivation of degree $\ell + \ell'$.

(iv) A derivation or a skew-derivation is completely determined by its effect on $\Lambda^0(M)$ and $\Lambda^1(M)$, that is, on the functions and 1-forms on M. Moreover, two derivations or skew-derivations D_1 and D_2 on $\Lambda(M)$ concide if and only if $D_1 f = D_2 f$, $D_1(df) = D_2(df)$, for every function f on M.

Notice that the Lie derivative $\mathcal{L}_X$ with respect to a vector field X on M is a derivation of degree 0 on $\Lambda(M)$ which commutes with the exterior differential d and the inner product i_X with respect to X is a skew-derivation of degree -1 on $\Lambda(M)$.

As relations among d, $\mathcal{L}_X$ and i_X, one have

PROPOSITION (8). (i) $[i_X,d] = i_X d + d i_X = \mathcal{L}_X$, for every vector field X. (ii) $[\mathcal{L}_X, i_Y] = i_{[X,Y]}$, for any vector fields X and Y.

REMARK (7). There is another approach to the theory of derivations due to Frölicher and Nijenhuis (1956). They define a derivation on $\Lambda(M)$ of degree ℓ as a linear mapping D of

$\Lambda(M)$ into itself which satisfies

$$\text{degree}\,(D\omega) = \text{degree}\,\omega + \ell$$

and

$$D(\omega\wedge\omega') = D\omega \wedge \omega' + (-1)^{p\ell}\omega + D\omega',$$

where $p = \text{degree}\,\omega$.

The exterior differential d is a derivation on $\Lambda(M)$ of degree 1. The commutator $[D,D'] = DD' - (-1)^{\ell\ell'}D'D$ of two derivations D and D' on $\Lambda(M)$ of degree ℓ and ℓ', respectively, is a derivation on $\Lambda(M)$ of degree $\ell + \ell'$.

Derivations on $\Lambda(M)$ are completely characterized by their action on $\Lambda^o(M)$ and $\Lambda^1(M)$ and being local mappings, they are characterized by their action on f and df, for each function f on M. Following Frölicher and Nijenhuis, we single out <u>derivations of type</u> i_* and <u>of type</u> d_*. A derivation D is of type i_* if it acts trivially on $\Lambda^o(M)$; D is a derivation of type d_* if $[D,d] = 0$. Moreover, they prove that every derivation on $\Lambda(M)$ can be decomposed in a unique way as a sum of a derivation of type i_* and a derivation of type d_*.

For each vector field X on M, there exists a derivation of type i_*, the inner product i_X, and a derivation of type d_*, the Lie derivative $\mathcal{L}_X$, canonically associated to X. If F is an endomorphism of TM, we can associate to F a derivation i_F of type i_* given by $(i_F\omega)(X_1,\ldots,X_p) = \sum_{i=1}^{p} \omega(X_1,\ldots,FX_i,\ldots,X_p)$, for any p-form ω and any vector fields $X_1,\ldots,X_p$ on M, and a derivation d_F of type d_* given by

$$d_F = [i_F,d] = i_Fd - di_F.$$

(For more details, we remit the reader to the quoted paper).

Let us return to the tangent bundle of order k, T^kM, of a manifold M. One can easily show the following

PROPOSITION (9). The mapping $\omega \to i_{J_r}\omega$, $1 \le r \le k$, is a derivation of degree 0 on the exterior algebra $\Lambda(T^kM)$, which will be called the "r th vertical derivation" on $\Lambda(T^kM)$.

This derivation i_{J_r} is completely determined by the following identities:

$$i_{J_r} f = 0$$
$$i_{J_r}(df) = J_r^*(df), \quad \text{for every function } f \text{ on } T^kM.$$

Then i_{J_r} is locally given by

$$\left.\begin{aligned} & i_{J_r} f = 0, \quad \text{for every function } f \text{ on } T^kM \\ & i_{J_r}(dz^A_{(i)}) = \begin{cases} 0 & , \text{ if } i < r \\ dz^A_{(i-r)} & , \text{ if } i \ge r \end{cases} \end{aligned}\right\} \tag{24}$$

where $0 \le i \le k$.

REMARK (8). Taking into account the theory of derivations developed by Frölicher and Nijenhuis, the derivation i_{J_r} is nothing but the derivation of type i_* associated to the endomorphism J_r of $T(T^kM)$.

PROPOSITION (10). We have

$$i_{J_r} \circ J_s^* = J_s^* \circ i_{J_r}.$$

Moreover,

$$i_{J_r} \circ J_s^* = J_s^* \circ i_{J_r} = 0, \quad \text{if} \quad r+s \ge k+1.$$

Proof: Let ω be a p-form and $X_1,\dots,X_p$ vector fields on

T^kM. One has

$$(i_{J_r} \circ J_s^*)(\omega)(X_1,\ldots,X_p) = \sum_{i=1}^{p} (J_s^*\omega)(X_1,\ldots,J_rX_i,\ldots,X_p)$$
$$= \sum_{i=1}^{p} \omega(J_sX_1,\ldots,J_sJ_rX_i,\ldots,J_sX_p)$$
$$= (i_{J_r}\omega)(J_sX_1,\ldots,J_sX_i,\ldots,J_sX_p)$$
$$= (J_s^* \circ i_{J_r})(\omega)(X_1,\ldots,X_p),$$

taking into account that $J_rJ_s = J_sJ_r$. Moreover, if $r+s \geq k+1$, one deduces that $J_rJ_s = 0$, which implies that $i_{J_r} \circ J_s^* = 0$. □

PROPOSITION (11). _We have_

(i) $[i_{J_r}, i_X] = i_{J_r}i_X - i_Xi_{J_r} = -i_{J_rX}$, _for every vector field_ X _on_ T^kM, $1 \leq r \leq k$,

(ii) $$[i_{J_r}, \mathcal{L}_{C_s}] = i_{J_r}\mathcal{L}_{C_s} - \mathcal{L}_{C_s}i_{J_r} = \begin{cases} 0 & , \text{ if } r+s > k+1 \\ s\, i_{J_{r+s-1}} & , \text{ if } r+s \leq k+1 \end{cases}$$

where $1 \leq r,s \leq k$.

Proof: It is sufficient to verify the equalities above in the following two cases: $\omega = f$, $\omega = df$, for every function f on T^kM. Then we have

$$[i_{J_r}, i_X]f = 0, \quad i_{J_rX}f = 0,$$
$$[i_{J_r}, i_X]df = -i_XJ_r^*(df) = -df(J_rX) = -i_{J_rX}(df),$$
$$[i_{J_r}, \mathcal{L}_{C_s}]f = 0,$$
$$[i_{J_r}, \mathcal{L}_{C_s}](df)(Y) = (i_{J_r}d(C_sf) - \mathcal{L}_{C_s}J_r^*(df))(Y)$$
$$= (J_rY)(C_sf) - C_s((J_rY)f) + (J_r[C_s,Y])f$$
$$= ([J_rY, C_s) + J_r[C_s,Y])f$$
$$= -([C_s, J_r]Y)f$$

$$= \begin{cases} 0 & , \text{ if } r+s > k+1 \\ s(J_{r+s-1}Y)f & , \text{ if } r+s \leq k+1, \end{cases}$$

taking into account Proposition (4).

On the other hand, we have

$$i_{J_{r+s-1}}(df)(Y) = (df)(J_{r+s-1}Y)$$

$$= (J_{r+s-1}Y)f, \quad \text{if} \quad r+s \leq k+1.$$

This ends the proof of the Proposition. ❑

COROLLARY (2). *We have*

$$[i_{J_r}, i_{C_s}] = \begin{cases} 0 & , \text{ if } r+s \geq k+1 \\ -i_{C_{r+s}} & , \text{ if } r+s < k+1, \end{cases}$$

where $1 \leq r,s \leq k$.

3.5. Vertical differentiation

The aim of this paragraph is to associate a skew-derivation of degree 1 on $\Lambda(T^kM)$ to each vertical endomorphism. To do this, we proceed as follows.

Let r be an integer, $1 \leq r \leq k$. Then we can consider the commutator $d_{J_r} = [i_{J_r}, d] = i_{J_r}d - di_{J_r}$. From (ii) of Proposition (7), we deduce that d_{J_r} is a skew-derivation of degree 1 on the exterior algebra $\Lambda(T^kM)$, which will be called the r th *vertical differentiation*.

The next proposition follows immediately from the definition of d_{J_r}.

PROPOSITION (12). *The* r th *vertical differentiation* d_{J_r} *is completely determined by the relations:*

$$d_{J_r} f = J_r^*(df),$$

$$d_{J_r}(df) = -d(J_r^* df),$$

for any function f on $T^k M$.

Therefore, d_{J_r} is locally given by

$$\left.\begin{aligned} d_{J_r} f &= \sum_{i=r}^{k} \frac{\partial f}{\partial z^A_{(i)}} dz^A_{(i-r)}, \\ d_{J_r}(dz^A_{(i)}) &= 0, \quad 0 \le i \le k, \end{aligned}\right\} \quad (25)$$

for any function f on $T^k M$.

We can particularize (24) for the cases $k = 1,2$. If we take $k = 1$, we have

$$d_J f = \frac{\partial f}{\partial z^A_{(1)}} dz^A_{(0)},$$

(or, $d_J f = \frac{\partial f}{\partial \dot{q}^A} dq^A$, in the usual notation for the local coordinates in TM).

If $k = 2$, we have

$$d_{J_1} f = \frac{\partial f}{\partial z^A_{(1)}} dz^A_{(0)} + \frac{\partial f}{\partial z^A_{(2)}} dz^A_{(1)},$$

$$d_{J_2} f = \frac{\partial f}{\partial z^A_{(2)}} dz^A_{(0)}.$$

PROPOSITION (13). *We have*

$$d_{J_r} d = -d d_{J_r}.$$

Proof: In fact, we have

$$d_{J_r} d = -d i_{J_r} d = -d d_{J_r}, \quad \text{since } d^2 = 0. \quad \square$$

REMARK (9). Notice that d_{J_r} is the derivation of type d_*

associated to the vertical endomorphism J_r in the terminology of Frölicher and Nijenhuis (1956).

PROPOSITION (14). <u>We have</u>

$$d_{J_r}^2 = 0, \qquad 1 \le r \le k.$$

<u>Proof</u>: It is sufficient to check that $d_{J_r}^2 \omega = 0$ in the cases $\omega = f$, $\omega = df$, f being an arbitrary function on T^kM. Let X, Y be two vector fields on T^kM. Then we have

$$\begin{aligned}
(d_{J_r}^2 f)(X,Y) &= d_{J_r}(J_r^* df)(X,Y) \\
&= (i_{J_r} d(J_r^* df))(X,Y) - (di_{J_r}(J_r^* df))(X,Y) \\
&= d(J_r^* df)(J_r X,Y) + d(J_r^* df)(X,J_r Y) \\
&\quad - X(i_{J_r}(J_r^* df)(Y)) + Y(i_{J_r}(J_r^* df)(X)) + (i_{J_r}(J_r^* df))[X,Y] \\
&= (J_r X)(J_r Y)f - Y(J_r^2 X)f - J_r[J_r X,Y]f \\
&\quad + X(J_r^2 Y)f - (J_r X)(J_r Y)f + J_r[X,J_r Y]f \\
&\quad - X(J_r^2 Y)f + Y(J_r^2 X)f + J_r^2[X,Y]f \\
&= ([J_r X,J_r Y] - J_r[J_r X,Y] - J_r[X,J_r Y] + J_r^2[X,Y])f \\
&= 0,
\end{aligned}$$

taking into account (iii) of Proposition (4). Finally, $d_{J_r}^2 df = dd_{J_r}^2 f = 0$. □

As relations among $\mathcal{L}_{C_r}$, i_{C_r}, i_{J_r}, d_{J_r}, $1 \le r \le k$, we have

PROPOSITION (15).

$$\text{(i)} \quad [i_{J_r}, d_{J_s}] = i_{J_r} d_{J_s} - d_{J_s} i_{J_r} = \begin{cases} 0, & \text{if } r+s \ge k+1 \\ d_{J_{r+s}}, & \text{if } r+s < k+1 \end{cases}$$

$$\text{(ii)}\quad [i_{C_r}, d_{J_s}] = i_{C_r} d_{J_s} + d_{J_s} i_{C_r} = \begin{cases} 0 & \text{, if } r+s > k+1 \\ (k-r+1)\, i_{J_k} & \text{, if } r+s = k+1 \\ \mathcal{L}_{C_{r+s}} + s\, i_{J_{r+s-1}} & \text{, if } r+s < k+1 \end{cases}$$

$$\text{(iii)}\quad [d_{J_r}, \mathcal{L}_{C_s}] = d_{J_r} \mathcal{L}_{C_s} - \mathcal{L}_{C_s} d_{J_r} = \begin{cases} 0 & \text{, if } r+s > k+1 \\ r\, d_{J_{r+s-1}} & \text{, if } r+s \leq k+1 \end{cases}$$

<u>Proof</u>: It is sufficient to verify the equalities above in the cases $\omega = f$, $\omega = df$, f being an arbitrary function on T^kM. Then, by the one hand, we have

$$\begin{aligned} [i_{J_r}, d_{J_s}] f &= i_{J_r} d_{J_s} f \\ &= i_{J_r} J_s^*(df) \\ &= \begin{cases} 0 \;, & \text{if } r+s \geq k+1 \\ J_s^* i_{J_r}(df) = J_s^* J_r^*(df), & \text{if } r+s < k+1. \end{cases} \end{aligned}$$

On the other hand, if $r+s < k+1$, we obtain

$$d_{J_{r+s}} f = J_{r+s}^*(df).$$

But $J_s^* J_r^* = J_{r+s}^*$, which shows that (i) is true for $\omega = f$.

Let us now take $\omega = df$. Then, by one hand, we have

$$\begin{aligned} [i_{J_r}, d_{J_s}] df &= -i_{J_r} d d_{J_s} f - d_{J_s} i_{J_r} df \\ &= -i_{J_r} d i_{J_s} df - i_{J_s} d i_{J_r} df + d i_{J_s} i_{J_r} df. \end{aligned}$$

Let X, Y be two vector fields on T^kM. Then, from a straightforward calculation, we deduce

$$\begin{aligned} [i_{J_r}, d_{J_s}](df)(X,Y) &= -[J_r, J_s](X,Y)f + X(J_r J_s Y)f \\ &\quad - Y(J_r J_s X)f + J_r J_s [X,Y] f \end{aligned}$$

$$= X(J_rJ_sY)f - Y(J_rJ_sX)f + J_rJ_s[X,Y]f.$$

On the other hand, if $r+s < k+1$, we have

$$d_{J_{r+s}}(df)(X,Y) = X(J_{r+s}Y)f - Y(J_{r+s}X)f - (J_{r+s}[X,Y])f.$$

Taking into account that

$$J_rJ_s = \begin{cases} 0\ , & \text{if} \quad r+s \geq k+1 \\ J_{r+s}\ , & \text{if} \quad r+s < k+1, \end{cases}$$

(i) is true also for $\omega = df$.

Formulas (ii) and (iii) can be proved in the same way as (i) and its verification is left to the reader as an exercise.

Before proceeding further, we will need the following lemma.

LEMMA. <u>We have</u>

$$J_r^* d_{J_s} f = \begin{cases} 0\ , & \text{if} \quad r+s > k \\ d_{J_{r+s}} f, & \text{if} \quad r+s \leq k, \end{cases}$$

for any function f on T^kM, where $1 \leq r,s \leq k$.

<u>Proof</u>: Indeed, from (25), we have

$$\begin{aligned} J_r^* d_{J_s} f &= J_s^*\Big(\sum_{i=s}^{k} \frac{\partial f}{\partial z^A_{(i)}} dz^A_{(i-s)}\Big) \\ &= \sum_{i=s}^{k} \frac{\partial f}{\partial z^A_{(i)}} J_r^*(dz^A_{(i-s)}) \\ &= \begin{cases} 0\ , & \text{if} \quad r+s > k \\ \displaystyle\sum_{i=r+s}^{k} \frac{\partial f}{\partial z^A_{(i)}} dz^A_{(i-s-r)}\ , & \text{if} \quad r+s \leq k \end{cases} \end{aligned}$$

$$= \begin{cases} 0 \quad , & \text{if} \quad r+s > k \\ d_{J_{r+s}} f \quad , & \text{if} \quad r+s \leq k. \quad \square \end{cases}$$

From this lemma, we can prove

PROPOSITION (16). We have

$$J_r^* d_{J_s} = 0, \quad \text{if} \quad r+s > k.$$

Proof: From the lemma, one has that $J_r^* d_{J_s} f = 0$, if $r+s > k$ and f is an arbitrary function on $T^k M$. Then, a simple calculation in local coordinates shows that $J_r^* d_{J_s} df = 0$, for any function f on $T^k M$. So, our proposition is proved. □

REMARK (10). If $r+s \leq k$, it is not true that $J_r^* d_{J_s} = d_{J_{r+s}}$. In fact, take $r = s = 1$, $k = 2$ and the function f on $T^2 M$ locally given by

$$f(z^A_{(0)}, z^A_{(1)}, z^A_{(2)}) = \frac{1}{2} \sum_A \{(z^A_{(0)})^2 + (z^A_{(2)})^2\}.$$

Then we have

$$J_1^* d_{J_1} df = \sum_A dz^A_{(0)} \wedge dz^A_{(1)}$$

and

$$d_{J_2} df = \sum_A dz^A_{(0)} \wedge dz^A_{(2)} .$$

To end this paragraph, we shall prove the following

PROPOSITION (17). We have

$$d_{J_r} J_r^* = J_r^* d,$$

where $1 \leq r \leq k$.

Proof: Let f be an arbitrary function on $T^k M$. Then we have

$$d_{J_r} J_r^* f = d_{J_r} f = J_r^*(df),$$

$$d_{J_r} J_r^*(df) = d_{J_r} d_{J_r} f = 0 = J_r^* d(df).$$

Now, if ω is a p-form on T^kM, we can easily deduce that $d_{J_r} J_r^* \omega = J_r^* d_{J_r} \omega$ by a direct calculation in local coordinates. Details are left to the reader. □

3.6. Semibasic forms

We shall now consider a special kind of differential forms on the tangent bundle of higher order.

DEFINITION (6). A differential form ω on T^kM is said to be semibasic of type r if $\omega \in \operatorname{Im} J_r^*$.

Let SB_r be the set of all semibasic forms on T^kM of type r. Then SB_r is a graduated subalgebra of the exterior algebra $\Lambda(T^kM)$, that is,

$$SB_r = \bigoplus_{p \geq 0} S^pB_r ,$$

$$(S^pB_r) \wedge (S^qB_r) \subset S^{p+q}B_r ,$$

where S^pB_r denotes the set of semibasic forms of degree p and type r.

Obviously, every function on T^kM is a semibasic 0-form of type r, $1 \leq r \leq k$, and every semibasic form of type s is, automatically, semibasic of type r, if $r \leq s$. Consequently, we have the following decreasing sequence of subalgebras:

$$SB_1 \supset SB_2 \supset \ldots \supset SB_k .$$

Let us now investigate the effect of the different derivations and skew-derivations on $\Lambda(T^kM)$ introduced in the previous paragraphs when they act on semibasic forms.

PROPOSITION (18). *We have*

(i) $d_{J_r}(SB_r) \subset SB_r$

(ii) $d_{J_r} f \in SB_r$, *for every function* f *on* T^kM.

Proof: (i) follows directly from Proposition (17), since $d_{J_r} \circ J_r^* = J_r^* \circ d$. (ii) is an immediate consequence of (i). □

PROPOSITION (19). *Let* r, s *be integers such that* $1 \leq r,s \leq k$. *Then* SB_r *is stable by the skew-derivation* i_{C_s}. *Particularly, if* $s \geq k-r+1$, *then* i_{C_s} *vanishes on* SB_r.

Proof: Let ω be a semibasic p-form of type r. Then $\omega = J_r^*\eta$, for some p-form η on T^kM. Therefore, we have

$$i_{C_s}\omega = i_{C_s}J_r^*\eta = \begin{cases} 0 , & \text{if} \quad r+s \geq k+1 \\ J_r^* i_{C_{r+s}}\eta , & \text{if} \quad r+s < k+1, \end{cases}$$

taking into account Corollary (1). □

PROPOSITION (20). SB_r *is stable by the action of* i_{J_s}, $1 \leq r,s \leq k$. *Particularly*, i_{J_s} *vanishes on* SB_r, *if* $s \geq k-r+1$.

Proof: Let ω be a p-form on T^kM. Then we have

$$i_{J_s} J_r^*\omega = \begin{cases} 0 , & \text{if} \quad r+s \geq k+1 \\ J_r^* i_{J_s}\omega , & \text{if} \quad r+s < k+1, \end{cases}$$

from Proposition (10). □

Let us take $k = 1$. Then these three last propositions can be summarized as follows: $d_J(SB) \subset SB$, $i_C(SB) = i_J(SB) = 0$ (see Godbillon (1969)). If $k = 2$, we have

$$SB_1 \supset SB_2, \quad d_{J_1}(SB_1) \subset SB_1, \quad d_{J_2}(SB_2) \subset SB_2,$$

$$i_{C_2}(SB_1) = i_{C_2}(SB_2) = i_{C_1}(SB_2) = 0, \quad i_{C_1}(SB_1) \subset SB_1,$$

$$i_{J_2}(SB_1) = i_{J_2}(SB_2) = i_{J_1}(SB_2) = 0, \quad i_{J_1}(SB_1) \subset SB_1.$$

Next, we shall obtain the local expression of a semibasic form. From (23), one easily deduces that the algebra SB_r is locally generated by the functions on T^kM and $dz^A_{(0)}, dz^A_{(1)}, \ldots, dz^A_{(k-r)}$. Therefore, a Pfaff form on T^kM is semibasic of type r if and only if it can be locally written as

$$\sum_0^{k-r} a_A^{(i)}(z^A_{(0)}, \ldots, z^A_{(k)})dz^A_{(i)}.$$

From these local expressions, it easily follows

PROPOSITION (21). Let α be a differential form on $T^{k-r}M$. Then $\beta = (\rho^k_{k-r})^*\alpha$ is a semibasic form of type r on T^kM such that $d_{J_{k-r+1}}\beta = 0$.

(Clearly, the terminology proposed in Definition (6) is now justified by the above Proposition).

In the rest of this paragraph, we will establish some properties of semibasic forms which will be useful in the Chapter II.

PROPOSITION (22). Let ω be a Pfaff form on T^kM. Then ω is semibasic of type r if and only if ω vanishes on the image of J_{k-r+1}.

Proof: In fact, $\operatorname{Im} J_{k-r+1} = V^{\rho^k_{k-r}}(T^kM)$. □

COROLLARY (3). Let ω be a Pfaff form on T^kM. Then ω is semibasic of type r if and only if there exists a function σ

on $T^kM \times_{T^{k-r}M} T(T^{k-r}M)$ such that:

(i) σ is linear on each fiber of $T^kM \times_{T^{k-r}M} T(T^{k-r}M)$

(ii) $\omega = \sigma \circ s_r$.

Proof: Our Corollary follows immediately taking into account Proposition (22) and the rth fundamental exact sequence:

$$0 \to V^{\rho^k_{k-r}}(T^kM) \xrightarrow{i_r} T(T^kM) \xrightarrow{s_r} T^kM \times_{T^{k-r}M} T(T^{k-r}M) \to 0. \quad \square$$

The Corollary (3) can be equivalently expressed as follows

COROLLARY (4). Let ω be a Pfaff form on T^kM. Then ω is semibasic of type r if and only if there exists a section σ of the vector bundle $(T^kM \times_{T^{k-r}M} T(T^{k-r}M))^*$ (dual of $T^kM \times_{T^{k-r}M} T(T^{k-r}M))$ such that

$$\omega(X) = \langle s_r(X), \sigma(p_{T^kM}(X))\rangle, \quad X \in T(T^kM).$$

Conversely, if σ is a section of the vector bundle $(T^kM \times_{T^{k-r}M} T(T^{k-r}M))^*$, then

$$\omega = \langle s_r, \sigma \circ p_{T^kM}\rangle$$

is a semibasic Pfaff form on T^kM of type r.

We adopt the following definition.

DEFINITION (7). The induced vector bundle of $T^*(T^{k-r}M) \xrightarrow{q_{T^{k-r}M}} T^{k-r}M$ via ρ^k_{k-r}, denoted by $T^kM \times_{T^{k-r}M} T^*(T^{k-r}M)$, will be called the bundle of semibasic forms of type r. (Here, $q_{T^{k-r}M}$ denotes the canonical projection onto $T^{k-r}M$ of the cotangent bundle $T^*(T^{k-r}M)$).

Then we have the following commutative diagram:

$$\begin{array}{ccc}
T^kM \times_{T^{k-r}M} T^*(T^{k-r}M) & \dashrightarrow & T^*(T^{k-r}M) \\
\Big\downarrow & & \Big\downarrow q_{T^{k-r}M} \\
T^kM & \xrightarrow{\rho^k_{k-r}} & T^{k-r}M
\end{array}$$

where the dotted arrows are the canonical projections.

Now, let h be the map from the Whitney sum $(T^kM \times_{T^{k-r}M} T(T^{k-r}M)) \oplus (T^kM \times_{T^{k-r}M} T^*(T^{k-r}M))$ into $\mathbb{R}$ given by $h((z,v),(z,\gamma)) = \langle v,\gamma\rangle$, where $\langle \quad \rangle$ denotes the natural pairing between vectors and 1-forms.

It is clear that the restriction of h to each fiber of $(T^kM \times_{T^{k-r}M} T(T^{k-r}M)) \oplus (T^kM \times_{T^{k-r}M} T^*(T^{k-r}M))$ is a non-degenerate bilinear form. Consequently, one easily deduces (see Godbillon (1969)).

PROPOSITION (23). *The vector bundle of semibasic forms of type* r *is equivalent to the dual vector bundle of* $T^kM \times_{T^{k-r}M} T(T^{k-r}M)$.

Hence, the Corollary (4) can be equivalently established as follows

THEOREM (1). *The correspondence*

$$\omega(X) = \langle T\rho^k_{k-r}(X),\ D(p_{T^kM}(X))\rangle, \quad X \in T(T^kM),$$

determines a bijection between the semibasic Pfaff forms of type r *on* T^kM *and the mappings* $D\colon T^kM \to T^*(T^{k-r}M)$ *such that*

$$q_{T^{k-r}M} \circ D = \rho^k_{k-r}\,.$$

Let us remark that if ω is locally given by

$$\omega = \sum_{i=0}^{k-r} a_A^{(i)} dz_{(i)}^A$$

and

$$X = \sum_{j=0}^{k} X_A^{(j)} \frac{\partial}{\partial z_{(j)}^A} ,$$

then we have

$$\omega(X) = \sum_{i=0}^{k-r} a_A^{(i)} X_A^{(i)} .$$

Therefore, the local expression of the function D corresponding to ω is

$$(z_{(0)}^A,\ldots,z_{(k)}^A) \rightsquigarrow (z_{(0)}^A,\ldots,z_{(k-r)}^A, a_A^{(0)},\ldots,a_A^{(k-r)}) \qquad (26)$$

Let us take $k = 1$, for a moment. According to (26), the function $D\colon TM \to T^*M$ corresponding to the semibasic Pfaff form $\omega = a_A dq^A$ is locally given by

$$(q^A, \dot{q}^A) \rightsquigarrow (q^A, a_A) \qquad (27)$$

As it is well known, there exists a canonical 1-form λ on the cotangent bundle T^*M of a manifold M given by

$$\lambda(\alpha)(Y) = \langle Tq_M(Y), \alpha \rangle, \quad \alpha \in T^*Q, \quad Y \in T_\alpha(T^*Q),$$

where $q_M\colon T^*M \to M$ is the canonical projection; λ is called the <u>Liouville form</u> on T^*M. Denoting by (q^A, p_A) the local coordinates in T^*M, then λ is locally given by

$$\lambda = p_A dq^A . \qquad (28)$$

From (27) and (28), we easily see that

$$D^*\lambda = \omega . \qquad (29)$$

We return now to the general case and obtain the following proposition which generalizes (29).

PROPOSITION (24). <u>If</u> λ_{k-r} <u>is the Liouville form on</u> $T^*(T^{k-r}M)$, <u>then we have</u>

$$D^*\lambda_{k-r} = \omega.$$

<u>Proof</u>: Directly from (26) and (28). □

3.7. Homogeneous vector fields and forms

The notion of homogeneity for functions on $\mathbb{R}^n$ can be extended in an obvious way for functions and differential forms on the tangent bundle (or, more generally, on the tangent bundles of higher order).

Indeed, let $h_t: \mathbb{R} \to \mathbb{R}$ be the homothetia of ratio e^t and $H_t: T^kM \to T^kM$ the fiber preserving transformation deduced from h_t. Then we can adopt the following.

DEFINITION (8). A function f on T^kM is said to be <u>homogeneous of degree</u> α if and only if $f \circ H_t = e^{\alpha t} \circ f$.

The next proposition gives a characterization of homogeneity of functions in terms of the Liouville vector field of higher order.

PROPOSITION (25). <u>A function</u> f <u>on</u> T^kM <u>is homogeneous of degree</u> α <u>if and only if</u>

$$\mathcal{L}_{C_1} f = \alpha f.$$

<u>Proof</u>: Suppose that f is homogeneous of degree α. Then we have

$$\mathcal{L}_{C_1} f = \lim_{t\to 0} \frac{H_t^* f - f}{t}$$

$$= \lim_{t\to 0} \frac{e^{\alpha t} f - f}{t}$$

$$= (\lim_{t\to 0} \frac{e^{\alpha t} - 1}{t}) f$$

$$= \alpha f,$$

since H_t is the flow generated by C_1 (Proposition (2)).

Conversely, suppose that $\mathcal{L}_{C_1} f = \alpha f$. Then, for every point $z \in T^k M$, $H_t^* f(z)$ is the solution of the differential equation $\frac{du}{dt} = \alpha u$ with initial condition $u(0) = f(z)$. From the uniqueness of solutions of differential equations, one deduces that $H_t^* f = e^{\alpha t} f$. □

The Proposition (25) can be used in order to obtain the local behavior of homogeneous functions. In fact, $\mathcal{L}_{C_1} f$ is locally given by

$$\mathcal{L}_{C_1} f = \sum_{i=1}^{k} i z_{(i)}^A \frac{\partial f}{\partial z_{(i)}^A} .$$

Then, f is homogeneous of degree α if and only if

$$\alpha f(z_{(0)}^A, z_{(1)}^A, \ldots, z_{(k)}^A) = \sum_{i=1}^{k} i z_{(i)}^A \frac{\partial f}{\partial z_{(i)}^A} . \tag{30}$$

The following definition is similar to Definition (8).

DEFINITION (9). A vector field X on $T^k M$ is said to be <u>homogeneous of degree</u> α if and only if

$$X \circ H_t = e^{(\alpha - 1)t}\, TH_t \circ X.$$

Note that if X is an homogeneous vector field of degree α on $T^k M$, we have the following commutative diagram:

$$\begin{array}{ccc} T(T^kM) & \xrightarrow{e^{(\alpha-1)t}TH_t} & T(T^kM) \\ \uparrow X & & \uparrow X \\ T^kM & \xrightarrow{H_t} & T^kM \end{array}$$

Let us suppose that $X = \sum_{i=0}^{k} X_A^{(i)} \frac{\partial}{\partial z_{(i)}^A}$ in local coordinates. Then an easy calculation shows that X is homogeneous of degree α if and only if the functions $X_A^{(i)}$ are homogeneous of degree $\alpha + i-1$, $0 \leq i \leq k$.

A useful criterion for the homogeneity of vector fields on T^kM is the following

PROPOSITION (26). <u>A vector field</u> X <u>on</u> T^kM <u>is homogeneous of degree</u> α <u>if and only if</u>

$$\mathcal{L}_{C_1}X = (\alpha-1)X.$$

The proof, which is similar to that of Proposition (25), is left to the reader as an exercise.

PROPOSITION (27). <u>Let</u> X_1 <u>and</u> X_2 <u>be homogeneous vector fields on</u> T^kM <u>of degree</u> α_1 <u>and</u> α_2, <u>respectively. Then</u> $[X_1,X_2]$ <u>is homogeneous of degree</u> $\alpha_1+\alpha_2-1$.

<u>Proof</u>: In fact, we have

$$\mathcal{L}_{C_1}[X_1,X_2] = -\mathcal{L}_{X_2}\mathcal{L}_{C_1}X_1 + \mathcal{L}_{X}\mathcal{L}_{C_1}X_2$$

(by virtue of Jacobi's identity)

$$= -(\alpha_1-1)\mathcal{L}_{X_2}X_1 + (\alpha_2-1)\mathcal{L}_{X_1}X_2$$

$$= (\alpha_1+\alpha_2-2)[X_1,X_2]. \quad \square$$

From Proposition (27), we see that the set of homogeneous vector fields on T^kM of degree 1 form a Lie subalgebra of the Lie algebra of all vector fields on T^kM.

Definition (8) can be extended for arbitrary differential forms as follows.

DEFINITION (10). A differential form ω on T^kM is said to be <u>homogeneous of degree</u> α if and only if

$$H_t^*\omega = e^{\alpha t}\omega.$$

We have

PROPOSITION (28). <u>A differential form</u> ω <u>on</u> T^kM <u>is homogeneous of degree</u> α <u>if and only if</u>

$$\mathcal{L}_{C_1}\omega = \alpha\omega.$$

Now, let $\omega = \sum_{i=0}^{k} a_A^{(i)} dz_{(i)}^A$ be the local expression of a Pfaff form on T^kM. Then we have

$$\mathcal{L}_{C_1}\omega = \sum_{i=1}^{k} i\, z_{(i)}^A \frac{\partial a_B^{(0)}}{\partial z_{(i)}^A} dz_{(0)}^B$$

$$+ \sum_{j=1}^{k} \left(\sum_{i=1}^{k} i\, z_{(i)}^A \frac{\partial a_B^{(j)}}{\partial z_{(i)}^A} + j \right) dz_{(j)}^B .$$

From (30) and Proposition (28), one easily deduces that ω is homogeneous of degree α if and only if $a_A^{(i)}$ is homogeneous of degree $\alpha-i$, $0 \leq i \leq k$.

Some properties of homogeneous vector fields and forms follow.

PROPOSITION (29). (i) <u>Let</u> ω <u>and</u> ω' <u>be homogeneous forms of degree</u> α <u>and</u> α', <u>respectively. Then</u> $\omega \wedge \omega'$ <u>is homogeneous of degree</u> $\alpha+\alpha'$.

(ii) Let X be an homogeneous vector field of degree α and f an homogeneous function of degree β. Then Xf is an homogeneous function of degree $\alpha+\beta-1$.

(iii) Let ω be an homogeneous p-form of degree α and $X_1,\ldots,X_p$ p homogeneous vector fields of degree β. Then $\omega(X_1,\ldots,X_p)$ is a homogeneous function of degree $\alpha+p(\beta-1)$.

PROPOSITION (30). Let ω be a homogeneous form on T^kM of degree α. Then $d\omega$, $i_{C_1}\omega$, $i_{J_1}\omega$ and $d_{J_1}\omega$ are homogeneous of degree α, α, $\alpha-1$ and $\alpha-1$, respectively.

Proof: In fact, we have

$$\mathcal{L}_{C_1} d\omega = d\mathcal{L}_{C_1}\omega = \alpha d\omega,$$

$$\begin{aligned}\mathcal{L}_{C_1} i_{C_1}\omega &= i_{C_1} d i_{C_1}\omega \\ &= i_{C_1}\mathcal{L}_{C_1}\omega \\ &= \alpha\, i_{C_1}\omega,\end{aligned}$$

$$\begin{aligned}\mathcal{L}_{C_1} i_{J_1}\omega &= i_{J_1}\mathcal{L}_{C_1}\omega - i_{J_1}\omega \\ &= \alpha\, i_{J_1}\omega - i_{J_1}\omega \\ &= (\alpha-1)\, i_{J_1}\omega\end{aligned}$$

(by virtue of Proposition (11),ii),

$$\begin{aligned}\mathcal{L}_{C_1} d_{J_1}\omega &= d_{J_1}\mathcal{L}_{C_1}\omega - d_{J_1}\omega \\ &= \alpha\, d_{J_1}\omega - d_{J_1}\omega \\ &= (\alpha-1)\, d_{J_1}\omega\end{aligned}$$

(by virtue of Proposition (15),iii). □

3.8. Semisprays and sprays of higher order

The aim of this section is to introduce a class of vector fiels on T^kM which will be very useful in the Generalized Classical Mechanics.

We first recall some well known geometrical facts about differential equations of first and second order. Let X be a vector field on an arbitrary manifold M. A curve σ in M satisfying the condition $\dot{\sigma} = X\circ\sigma$ is called an integral curve of X. Suppose that σ and X are locally given in a chart by $\sigma = (\sigma^1,\ldots,\sigma^m)$, $X = \sum_{A=1}^{m} X^A \frac{\partial}{\partial x^A}$, $m = \dim M$, respectively. Then σ is an integral curve of X if and only if it verifies

$$\frac{d\sigma^A}{dt} = X^A(\sigma^1,\ldots,\sigma^m), \quad 1 \leq A \leq m. \tag{31}$$

Now, we shall consider differential equations which arise from a certain class of vector fields ξ on TM characterized as follows: Every integral curve σ of ξ satisfies the condition

$$\widehat{p_M \circ \sigma}^{\,\bullet} = \sigma.$$

Such a vector field ξ will be called a semispray (semigerbe, in French) and it is locally given by

$$\xi = \dot{q}^A \frac{\partial}{\partial q^A} + \xi^A(q,\dot{q}) \frac{\partial}{\partial \dot{q}^A}. \tag{32}$$

From (32), one easily deduces that a vector field ξ on TM (that is, a section of the tangent bundle $T(TM) \xrightarrow{p_{TM}} TM$ of TM) is a semispray if and only if $Tp_M\circ\xi = Id_{TM}$, that is, ξ is also a section of the vector bundle $T(TM) \xrightarrow{Tp_M} TM$ (see Godbillon (1969)).

If $\dot{\sigma}$ is the canonical lift to TM of a curve σ in M, we denote by $\ddot{\sigma}$ the canonical lift of $\dot{\sigma}$ to TTM. A curve σ in M is called a solution (or path) of ξ, if and only if $\dot{\sigma}$ is an integral curve of ξ, that is, σ verifies $\ddot{\sigma} = \xi \circ \dot{\sigma}$. Taking into account the local expressions of σ and ξ, one deduces that σ is a solution of ξ if and only if it satisfies the following system of differential equations of second order:

$$\frac{d^2\sigma^A}{dt^2} = \xi^A(\sigma^1,\ldots,\sigma^m, \frac{d\sigma^1}{dt},\ldots,\frac{d\sigma^m}{dt}), \tag{33}$$

$1 \leq A \leq m$.

Now, we shall consider the higher order case, and, on the sight of the previous discussion it is natural to adopt the following

DEFINITION (11). A vector field ξ on T^kM is said to be a semispray of type r, $1 \leq r \leq k$, if and only if

$$\widetilde{(\beta^k \circ \sigma)}^{k-r+1} = \rho^k_{k-r+1} \circ \sigma,$$

for every integral curve σ of ξ, where $\widetilde{(\beta^k \circ \sigma)}^{k-r+1}$ denotes the canonical lift of $\beta^k \circ \sigma$ to $T^{k-r+1}M$. (For $r = 1$, we put $\rho^k_k = Id_{T^kM}$).

The following diagram illustrates the definition above:

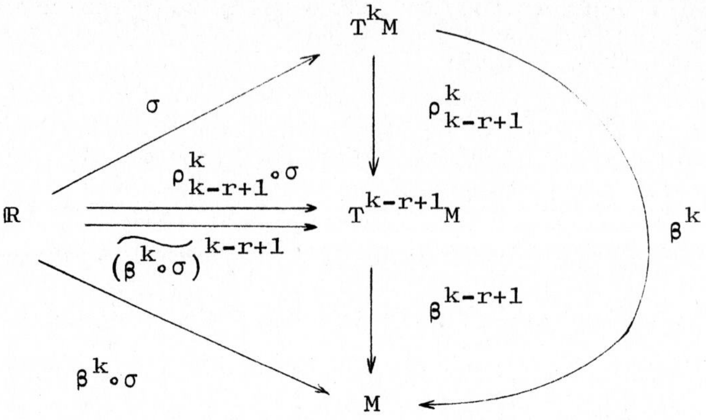

One can easily deduce that a semispray ξ on T^kM of type r is locally given by

$$\xi(z^A_{(0)},z^A_{(1)},\ldots,z^A_{(k)}) = (z^A_{(0)},z^A_{(1)},\ldots,z^A_{(k)};z^A_{(1)},2z^A_{(2)},\ldots,$$
$$\ldots,(k-r+1)z^A_{(k-r+1)},z^A_{(k-r+1)},\ldots,\xi^A_{(k)}),$$

or, equivalently,

$$\xi = z^A_{(1)}\frac{\partial}{\partial z^A_{(0)}} + 2z^A_{(2)}\frac{\partial}{\partial z^A_{(1)}} + \ldots + (k-r+1)z^A_{(k-r+1)}\frac{\partial}{\partial z^A_{(k-r)}}$$
$$+ \xi^A_{(k-r+1)}\frac{\partial}{\partial z^A_{(k-r+1)}} + \ldots + \xi^A_{(k)}\frac{\partial}{\partial z^A_{(k)}} \qquad (34)$$

From (13), §3.2, and (34), one has that a vector field ξ on T^kM is a semispray of type r if and only if the following diagram

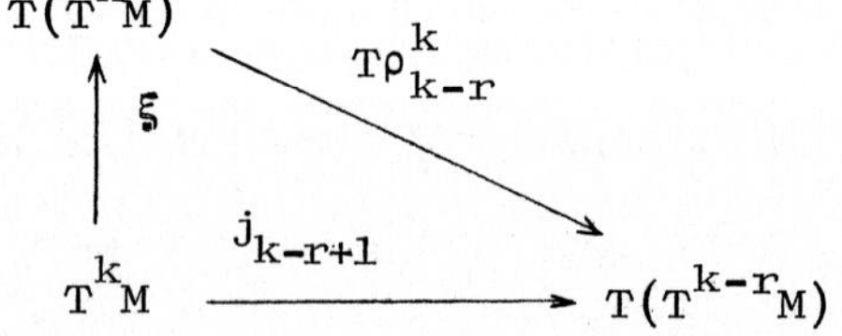

is commutative, that is,

$$T\rho^k_{k-r}\circ\xi = j_{k-r+1}\,.$$

It is convenient, for later use, to particularize (34) for $k = 2$. Then, we obtain

$$\xi = z^A_{(1)} \frac{\partial}{\partial z^A_{(0)}} + 2z^A_{(2)} \frac{\partial}{\partial z^A_{(1)}} + \xi^A \frac{\partial}{\partial z^A_{(2)}}, \qquad (35)_1$$

if ξ is of type 1, and

$$\xi = z^A_{(1)} \frac{\partial}{\partial z^A_{(0)}} + \xi^A \frac{\partial}{\partial z^A_{(1)}} + \bar{\xi}^A \frac{\partial}{\partial z^A_{(2)}}, \qquad (35)_2$$

if ξ is of type 2.

Therefore, a vector field ξ on T^2M is a semispray of type 1 (resp., of type 2) if and only if the following diagram

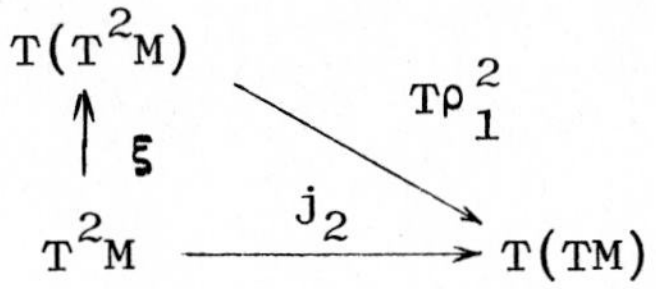

(resp.,

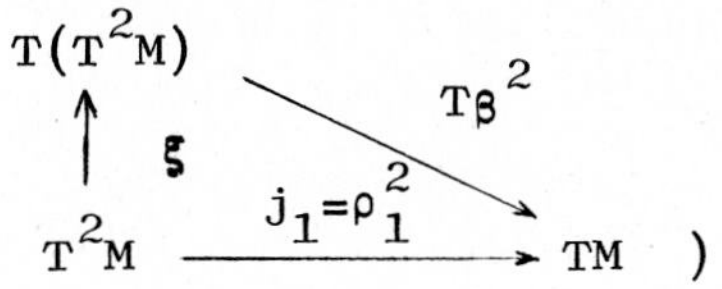

is commutative.

From (34), we have

PROPOSITION (31). <u>A vector field</u> ξ <u>on</u> T^kM <u>is a semispray of type</u> r <u>if and only if</u> $J_r\xi = C_r$.

From Proposition (31), one can show that a semispray of type r is of type s, for every integer s, $s \geq r$.

DEFINITION (12). Let ξ be a semispray on T^kM of type r. A curve σ in M is called a <u>path</u> (or <u>solution</u>) of ξ if and only if $\tilde{\sigma}^k$ is an integral curve of ξ, that is,

$$\dot{\widehat{\tilde{\sigma}^k}} = \xi \circ \tilde{\sigma}^k.$$

Consequently, a curve σ in M is a path of ξ if and only if it satisfies the following system of differential equations of order $k+1$:

$$\left.\begin{array}{l} \frac{1}{(k-r+1)!}\frac{d^{k-r+2}\sigma^A}{dt^{k-r+2}} = \xi^A_{(k-r+1)}\left(\sigma^A, \frac{d\sigma^A}{dt}, \ldots, \frac{d^k\sigma^A}{dt^k}\right) \\ \cdots\cdots\cdots\cdots\cdots\cdots \\ \frac{1}{k!}\frac{d^{k+1}\sigma^A}{dt^{k+2}} = \xi^A_{(k)}\left(\sigma^A, \frac{d\sigma^A}{dt}, \ldots, \frac{d^k\sigma^A}{dt^k}\right) \end{array}\right\} \quad (36)$$

which generalizes (31) and (33).

In the case $k = 2$, (36) becomes

$$\frac{1}{2}\frac{d^3\sigma^A}{dt^3} = \xi_A\left(\sigma^A, \frac{d\sigma^A}{dt}, \frac{d^2\sigma^A}{dt^2}\right) \quad (37)_1$$

if ξ is of type 1, and

$$\left.\begin{array}{l} \frac{d^2\sigma^A}{dt^2} = \xi^A\left(\sigma^A, \frac{d\sigma^A}{dt}, \frac{d^2\sigma^A}{dt^2}\right) \\ \frac{d^3\sigma^A}{dt^2} = \bar{\xi}^A\left(\sigma^A, \frac{d\sigma^A}{dt}, \frac{d^2\sigma^A}{dt^2}\right) \end{array}\right. \quad (37)_2$$

if ξ is of type 2.

We now describe a special type of semispray which will play a fundamental role in Classical Mechanics. As we know, if σ is an integral curve of a vector field ξ on M, then so is $\sigma\circ\gamma$, where $\gamma: \mathbb{R} \to \mathbb{R}$ is any translation. A similar result is true for a semispray ξ on TM. We now consider more general changes of parameter $\gamma: \mathbb{R} \to \mathbb{R}$ of the form $t \mapsto at+b$, where $a > 0$. The semispray ξ in which we are interested are characterized by the property that if σ is any path then so is $\sigma \circ \gamma$: such semispray ξ is called a <u>spray</u> on TM and its paths are called the <u>geodesics</u> of the spray.

This restriction imposes conditions on the vector field ξ. Suppose that the semispray ξ is locally given by (32). Then we can easily prove (see, for instance, the book of Brickell and Clark (1970)) that ξ <u>is a spray if and only if the functions</u> ξ^A <u>are homogeneous of degree</u> 2. Taking into account the results obtained in §3.6, one deduces that <u>a semispray</u> ξ <u>is a spray if and only if</u> $[C,\xi] = \xi$, <u>that is</u>, ξ <u>is a homogeneous vector field on</u> TM <u>of degree</u> 2. This justifies the following definition.

DEFINITION (13). A semispray ξ on T^kM of type r is called a <u>spray of type</u> r if and only if

$$[C_1,\xi] = \xi,$$

that is, ξ is homogeneous of degree 2. If ξ is a spray, its paths are called the <u>geodesics</u> of the spray.

From $(35)_1$ and $(35)_2$, one deduces that a semispray ξ on T^2M of type 1 is a spray if and only if the functions ξ^A are homogeneous of degree 3, and a semispray ξ on T^2M of type 2 is a spray if and only if the functions ξ^A (resp., $\bar{\xi}^A$) are homogeneous of degree 2 (resp., 3). For an arbitrary k, we deduce, from (34), that a semispray ξ on T^kM of type r is a spray if and only if the functions $\xi^A_{(j)}$, $r \leq j \leq k$, are homogeneous of degree $j+1$.

REMARK (11). The notion of spray has been introduced by Ambrose, Palais and Singer (1960) and was generalized by

Dazord (1966). An excelent reference on this subject is the book of Brickell & Clark (1970) or Grifone paper (1972).

3.9. Connections of higher order

Our aim in this section is to get a characterization of the connections on tangent bundles of higher order by means of the canonical almost tangent structures of higher order and develop a similar study to the one elaborated by Grifone (1972) in the case $k = 1$. The results obtained in this section will be useful in the Chapter II.

Let $p: E \to M$ be a fibered manifold and denote by $\text{Ver}\, E$ the vertical bundle of E over M. By a <u>connection</u> on E we mean a vector bundle H over E such that

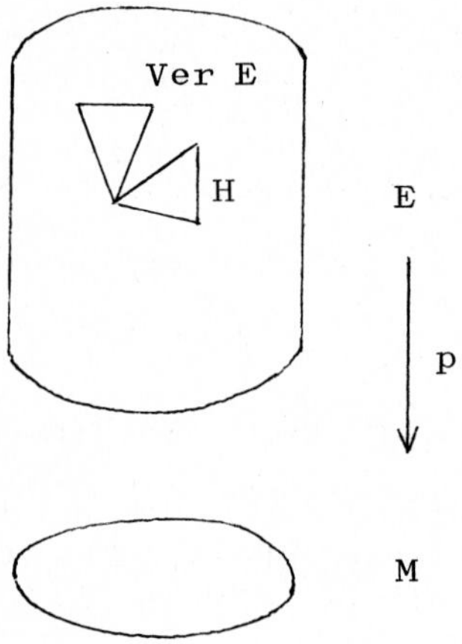

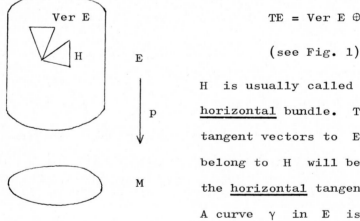

$$TE = \text{Ver}\, E \oplus H \qquad (38)$$

(see Fig. 1)

H is usually called the <u>horizontal</u> bundle. The tangent vectors to E which belong to H will be called the <u>horizontal</u> tangent vectors. A curve γ in E is said to be <u>horizontal</u> if all its tangent vectors are horizontal.

We now consider the tangent bundle $p_M: TM \to M$ of M and suppose a connection on TM with horizontal bundle H is given. Since $\text{Ver}(TM) = \text{Im}\, J = \text{Ker}\, J$, J being the canonical almost tangent structure of TM, it is easy to check that $Jh = h$, $hJ = 0$ and $Jv = 0$, $vJ = J$, where h and v are the projection operators $h\colon T(TM) \to H$, $v\colon T(TM) \to V(TM)$ determined by (38).

Then, if we put $\Gamma = 2h - Id$, we have that Γ is an endomorphism of $T(TM)$ such that

$$J\Gamma = J, \quad \Gamma J = -J. \tag{39}$$

Conversely, suppose that Γ is an endomorphism of $T(TM)$ such that (39) is verified; then $\Gamma^2 = Id$ and, if we put $h = \frac{1}{2}(Id+\Gamma)$, $v = \frac{1}{2}(Id-\Gamma)$, we have $\operatorname{Im} v = V(TM)$ and $H = \operatorname{Im} h$ defines a connection on TM.

So, we can give the following definition.

DEFINITION (14). An endomorphism Γ of $T(TM)$ satisfying (39) will be called a <u>connection on</u> TM or, simply, a <u>connection on</u> M (Grifone (1972)).

Not, let T^kM be the tangent bundle of order k of M. For each integer r, $1 \leq r \leq k$, T^kM is fibered over $T^{r-1}M$. Hence, we can consider connections on $T^kM \xrightarrow{\rho^k_{r-1}} T^{r-1}M$, for each r, and, taking into account (21), one can characterize this kind of connections as follows.

DEFINITION (15). An endomorphism Γ of $T(T^kM)$ such that

$$J_r\Gamma = J_r, \quad \Gamma J_{k-r+1} = -J_{k-r+1}, \tag{40}$$

will be called a <u>connection on</u> T^kM <u>of type</u> r, or simply, a <u>connection on</u> M <u>of order</u> k <u>and type</u> r.

If Γ is a connection on T^kM of type r, we denote by $h = \frac{1}{2}(Id+\Gamma)$, $v = \frac{1}{2}(Id-\Gamma)$ the projection operators associated to Γ, which will be called the horizontal and vertical projection operators of Γ, respectively. Then we have

$$h+v = Id,$$

$$J_r h = J_r, \quad h\, J_{k-r+1} = 0, \quad J_r v = 0, \quad v\, J_{k-r+1} = J_{k-r+1}.$$

It is easy to prove that the associated matrix to a connection Γ on T^kM of type r, in local coordinates, is

$$\Gamma = \begin{pmatrix} I_{(k-r+1)m} & 0 \\ -2\Gamma^{(\lambda,s)} & -I_{rm} \end{pmatrix} \tag{41}$$

where $\Gamma^{(\lambda,s)}$ is a matrix $m\times m$, $0 \leq \lambda \leq r-1$, $0 \leq s \leq k-r$, $m = \dim M$.

If we take $k = 1$, (41) becomes simply

$$\Gamma = \begin{pmatrix} I_m & 0 \\ -2\Gamma^B_A & -I_m \end{pmatrix} \tag{42}$$

and, if $k = 2$, (41) becomes

$$\Gamma = \begin{pmatrix} I_m & 0 & 0 \\ 0 & I_m & 0 \\ -2\Gamma^B_A & -2\bar{\Gamma}^B_A & -I_m \end{pmatrix} \tag{43$_1$}$$

if Γ is of type 1, and

$$\Gamma = \begin{pmatrix} I_m & 0 & 0 \\ -2\Gamma^B_A & -I_m & 0 \\ -2\bar{\Gamma}^B_A & 0 & -I_m \end{pmatrix} \tag{43$_2$}$$

if Γ is of type 2.

Now, we are interested in a special kind of connections on T^kM; namely, those connections whose horizontal bundle is invariant by the action of the homothetias of T^kM.

DEFINITION (16). Let Γ be a connection on T^kM of type r. The endomorphism H of TT^kM given by

$$H = \frac{1}{2}[C_1,\Gamma],$$

that is,

$$H(X) = \frac{1}{2}([C_1,\Gamma X] - \Gamma[C_1,X]),$$

for any vector field X on T^kM, will be called the <u>tension</u> of Γ. Then, a connection Γ will be said <u>homogeneous</u> if its tension H vanishes.

A straightforward calculation in local coordinates shows that if the associated matrix to Γ is given by (41), then the associated matrix to H is given by

$$H = \begin{pmatrix} 0 & 0 \\ & \\ H^{(\lambda,s)} & 0 \end{pmatrix}, \tag{44}$$

where $H^{(\lambda,s)} = -\sum_{j=1}^{k} j\, z^A_{(j)} \dfrac{\partial\Gamma^{(\lambda,s)}}{\partial z^A_{(j)}} + (k-r+1-\lambda-s)\Gamma^{(\lambda,s)}$,

$0 \leq \lambda \leq r-1$, $0 \leq s \leq k-r$.

The reader can easily check from (42), $(43)_1$ and $(43)_2$ that (44) becomes

$$H = \begin{pmatrix} 0 & 0 \\ & \\ \Gamma^B_A - \dot{q}^c \dfrac{\partial\Gamma^B_A}{\partial\dot{q}^c} & 0 \end{pmatrix} \tag{45}$$

in the case $k = 1$, and, for $k = 2$,

$$H = \begin{pmatrix} 0 & 0 & 0 \\ 0 & 0 & 0 \\ 2\Gamma^B_A - z^c_{(1)} \frac{\partial \Gamma^B_A}{\partial z^c_{(1)}} - 2z^c_{(2)} \frac{\partial \Gamma^B_A}{\partial z^c_{(2)}} & \bar{\Gamma}^B_A - z^c_{(1)} \frac{\partial \bar{\Gamma}^B_A}{\partial z^c_{(1)}} - 2z^c_{(2)} \frac{\partial \bar{\Gamma}^B_A}{\partial z^c_{(2)}} & 0 \end{pmatrix} \quad (46)_1$$

if Γ is of type 1, and

$$H = \begin{pmatrix} 0 & 0 & 0 \\ \Gamma^B_A - z^c_{(1)} \frac{\partial \Gamma^B_A}{\partial z^c_{(1)}} - 2z^c_{(2)} \frac{\partial \Gamma^B_A}{\partial z^c_{(2)}} & 0 & 0 \\ 2\bar{\Gamma}^B_A - z^c_{(1)} \frac{\partial \bar{\Gamma}^B_A}{\partial z^c_{(1)}} - 2z^c_{(2)} \frac{\partial \bar{\Gamma}^B_A}{\partial z^c_{(2)}} & 0 & 0 \end{pmatrix} \quad (46)_2$$

if Γ is of type 2.

From (44), it follows that Γ is an homogeneous connection if and only if the functions $\Gamma^{(\lambda,s)}$ are homogeneous of degree $k-r+1-\lambda-s$ and, hence, Γ is homogeneous if and only if the horizontal bundle defined by Γ is invariant by the action of the homothetias.

For instance, if Γ is a connection on T^2M of type 1, $(46)_1$ shows that Γ is homogeneous if and only if Γ^B_A and $\bar{\Gamma}^B_A$ are homogeneous of degree 2 and 1, respectively. If Γ is a connection of type 2, $(46)_2$ shows that Γ is homogeneous if and only if Γ^B_A and $\bar{\Gamma}^B_A$ are homogeneous of degree 1 and 2 respectively.

Let us return, for a moment, to the case $k = 1$. We recall that, for every $z \in T_xM$, $x \in M$, there is a linear isomorphism

$$V_z(TM) \xrightarrow{L_z} T_xM$$

locally defined by

$$\sum_{A=1}^{m} a^A \frac{\partial}{\partial \dot{q}^A} \rightsquigarrow \sum_{A=1}^{m} a^A \frac{\partial}{\partial q^A} \qquad \text{(Grifone (1972)).}$$

Let X, Y be vector fields on M and suppose a connection Γ on TM is given.

Then we can define a new vector field $\nabla_X Y$ on M as follows:

$$(\nabla_X Y)(x) = (L_{Y(x)} \circ v \circ dY(x))X(x),$$

for every $x \in M$, where $v = \frac{1}{2}(Id-\Gamma)$.

If $X = X^A \frac{\partial}{\partial q^A}$, $Y = Y^A \frac{\partial}{\partial q^A}$ are the local expressions of the vector fields X and Y, respectively, we deduce, from (42), that

$$\nabla_X Y = X^A \left(\frac{\partial Y^B}{\partial q^A} + \Gamma^B_A(q,Y)\right)\frac{\partial}{\partial q^B}. \tag{47}$$

If, in addition, Γ is homogeneous, we deduce, from (45), that the functions Γ^B_A are homogeneous of degree 1, and, moreover, linear on $\dot{q}^C$. Then we can put

$$\Gamma^B_A(q,\dot{q}) = \dot{q}^C \, \Gamma^B_{AC}(q),$$

that is,

$$\Gamma^B_{AC}(q) = \frac{\partial \Gamma^B_A}{\partial \dot{q}^C}.$$

Hence, from (47), we deduce that ∇ verifies the following conditions:

$$\nabla_{X+X'}Y = \nabla_X Y + \nabla_{X'}Y, \quad \nabla_Y(Y+Y') = \nabla_X Y + \nabla_X Y'$$

$$\nabla_{fX}Y = f\nabla_X Y, \quad \nabla_X(fY) = (\alpha f)Y + f\nabla_X Y,$$

for any vector fields X, X', Y, Y' and any function f on M. Then ∇ is a linear connection on M (see Koszul (1960)). Conversely, a linear connection ∇ on M defines a homogeneous connection Γ on TM (Grifone (1972)).

REMARK (12). If, in the Definition (14), we restrict ourselves to the tangent bundle without the zero section, then we can distinguish between linear and strictly homogeneous connections (see Vilms (1967) and Grifone (1972)).

Next, we define the geodesics of higher order connections.

DEFINITION (17). Let Γ be a connection on T^kM of type r and σ a curve in M. Then σ will be called a path of Γ if $\tilde{\sigma}^k$ is a horizontal curve in T^kM. If Γ is homogeneous, then the paths of Γ will be called the geodesics of Γ.

Let us remark that $\tilde{\sigma}^k$ is horizontal if and only if $v(\dot{\widehat{\tilde{\sigma}^k}}) = 0$. Hence, a curve σ in M is a path of a connection Γ if and only if σ satisfies the following system of differential equations:

$$\frac{1}{\lambda!}\frac{d^{\lambda+1}z^B(0)}{dt^{\lambda+1}} = -\sum_{s=0}^{k-r}\frac{1}{s!}\Gamma^{(\lambda-k+r-1,s)B}_{A}\frac{d^{s+1}z^A(0)}{dt^{s+1}}, \qquad (48)$$

where $k-r+1 \leq \lambda \leq k$.

For $k = 1$, (48) becomes simply

$$\frac{d^2q^B}{dt^2} = -\Gamma^B_A\frac{dq^A}{dt}, \quad 1 \leq B \leq m \qquad (49)$$

and, if Γ is homogeneous, (49) can be written as follows

$$\frac{d^2q^B}{dt^2} = -\Gamma^B_{AC}\frac{dq^A}{dt}\frac{dq^C}{dt}, \qquad 1 \le B \le m.$$

If Γ is a connection on T^2M, then (48) becomes

$$\frac{1}{2}\frac{d^3z^B_{(0)}}{dt^3} = -\Gamma^B_A\frac{dz^A_{(0)}}{dt} - \bar{\Gamma}^B_A\frac{d^2z^A_{(0)}}{dt^2}, \qquad (50)_1$$

if Γ is of type 1, and

$$\left.\begin{aligned}\frac{d^2z^B_{(0)}}{dt^2} &= -\Gamma^B_A\frac{dz^A_{(0)}}{dt}\\ \frac{1}{2}\frac{d^3z^B_{(0)}}{dt^3} &= -\bar{\Gamma}^B_A\frac{dz^A_{(0)}}{dt}\end{aligned}\right\} \qquad (50)_2$$

if Γ is of type 2.

Now, let Γ be a connection on T^kM of type r. If ξ' is an arbitrary semispray on T^kM of type r, then $\xi = h\xi'$ is again a semispray on T^kM of type r. Indeed, $J_r\xi = J_rh\xi' = J_r\xi' = C_r$. Moreover, ξ does not depend on the choice of ξ', since if ξ'' is another semispray on T^kM of type r, then $\xi'-\xi''$ is a vertical vector field and, consequently,

$$h(\xi'-\xi'') = h\xi' - h\xi'' = 0.$$

Next, we will obtain the local expression of ξ. From (41), we easily deduce that

$$\xi = (z^A_{(1)}, 2z^A_{(2)}, \ldots, (k-r+1)z^A_{(k-r+1)}, \xi^A_{(k-r+2)}, \ldots, \xi^A_{(k)}),$$

where

$$\xi^A_{(\lambda)} = -\sum_{s=0}^{k-r}(s+1)z^B_{(s+1)}\,\Gamma^{(\lambda-k+r-1,s)A}_{\quad B}, \qquad (51)$$

$k-r+1 \le \lambda \le k$.

If we particularize (48) for $k = 1$, one has

$$\xi^A = -\dot{q}^B \Gamma^A_B , \tag{52}$$

or,

$$\xi^A = -\dot{q}^B \dot{q}^C \Gamma^A_{BC} ,$$

if Γ is homogeneous.

If we take $k = 2$, one deduces

$$\xi^A = -z^B_{(1)} \Gamma^A_B - 2z^B_{(2)} \bar{\Gamma}^A_B , \tag{53$_1$}$$

if Γ is of type 1, and

$$\left.\begin{aligned} \xi^A &= -z^B_{(1)} \Gamma^A_B \\ \bar{\xi}^A_{(2)} &= -z^B_{(1)} \bar{\Gamma}^A_B \end{aligned}\right\} \tag{53$_2$}$$

if Γ is of type 2.

From (51), the reader can easily check that if Γ is homogeenous, then ξ is a spray (the calculus becomes very simple if we consider $k = 1$, and use (52), or $k = 2$, and use $(53)_1$ and $(53)_2$).

Summarizing the discussion above, we have

PROPOSITION (32). <u>There exists a semispray</u> ξ <u>on</u> T^kM <u>of type</u> r <u>canonically associated to each connection</u> Γ <u>on</u> T^kM <u>of type</u> r <u>in such a way that if</u> Γ <u>is homogeneous then its associated semispray</u> ξ <u>is a spray.</u>

Now, from (36) and (48), one deduces

PROPOSITION (33). <u>A connection</u> Γ <u>on</u> T^kM <u>of type</u> r <u>and its associated semispray</u> ξ <u>have the same paths.</u>

To end this section, we consider the following question: "Given a semispray ξ on T^kM of type r, there is some connection Γ on T^kM of type r such that the semispray associated to Γ is precisely ξ?" We answer affirmatively for the homogeneous case and $r = 1$.

PROPOSITION (34). *Let* ξ *be a spray on* T^kM *of type* 1. *If we put*

$$\Gamma = \frac{1}{k+1}\{-2[\xi, J_1] + (k-1)Id\},$$

then we have:

(i) Γ *is a homogeneous connection on* T^kM *of type* 1

(ii) *the spray associated to* Γ *is precisely* ξ.

I.4. - Exercises

1. Let M be an m-dimensional manifold and T^2M the tangent bundle of order 2 of M. Let us denote

$$\tilde{T}_2M = \{v;\ v \in TTM \text{ and } Iv = v\},$$

where I is the canonical involution of TTM (see Godbillon (1969)).

(i) Prove that the mapping $j_2\colon T^2M \to TTM$, given by (13), § 3.2, defines a diffeomorphism between T^2M and $\tilde{T}_2M$.

(ii) Let $\xi\colon T^2M \to T^3M$ be a mapping and put $\tilde{\xi} = j_3 \circ \xi$, where $j_3\colon T^3M \to TT^2M$. Prove that $\tilde{\xi}$ is a semispray on T^2M of type 1 if and only if ξ is a section of $T^3M \xrightarrow{\rho^3_2} T^2M$.

(iii) Extend (i) and (ii) for an arbitrary k (see Rodrigues (1975)).

2. Prove the parts (ii) and (iii) of Proposition (15), §3.5.

3. Prove Proposition (26), §3.7.

4. Prove that if ω is a p-form on T^kM homogeneous of degree α and r an integer, $1 \leq r \leq k$, then $i_{J_r}\omega$ and $d_{J_r}\omega$ are also homogeneous of degree $\alpha-1$ and $\alpha-r$, respectively.

5. Let $p\colon E \to M$ be a fibered manifold and denote by VE the vertical bundle over E. Then we can consider an exact sequence of vector bundles over E:

$$0 \to VE \xrightarrow{i} TE \xrightarrow{s} E \times_M TM \to 0 \tag{1}$$

(i) Prove that a connection on E determines a left splitting of (1) and, conversely, each left splitting of (1) defines a connection on E (see Vilms (1967)).

(ii) Prove that a left splitting of the (k-r+1)-th fundamental exact sequence defines a connection of type r on T^kM (see Catz $(1974)_b$ and de León (1981) for k=2 and de León & Villaverde (1981) for an arbitrary k).

6. Let TM be the tangent bundle of a manifold M. An endomorphism F of TM is said to be an almost complex structure on M if $F^2 = -Id$, Id being the identity endomorphism of TM. Suppose now that there is a Riemannian metric $\langle\ ,\ \rangle$ on M such that $\langle FX,FY\rangle = \langle X,Y\rangle$, for any vector fields X, Y on M; then $\langle\ ,\ \rangle$ is called <u>Hermitian metric</u> with respect to F and $(F,\langle\ ,\ \rangle)$ an <u>almost Hermitian structure</u> on M. Then we can define a 2-form Ω on M given by $\Omega(X,Y) = \langle FX,Y\rangle$, for any vector fields X, Y on M;

Ω will be called the Kaehler form of the almost Hermitian structure $(F, \langle\ ,\ \rangle)$.

Suppose now that Γ is a connection on TM with horizontal and vertical projection operators h and r, respectively, and define an endomorphism F of TTM by

$$Fv = h \quad \text{and} \quad Fh = -v.$$

(i) Prove that F is an almost complex structure on TM (see Grifone (1972)).

(ii) Let $(M, \langle\ ,\ \rangle)$ be a Riemannian manifold and Γ the Levi-civita connection determined by $\langle\ ,\ \rangle$. We define a Riemannian metric $\langle\ ,\ \rangle'$ on TM given by

$$\langle hX, hY\rangle'_z = \langle Tp_M X,\ Tp_M Y\rangle_{p_M(z)},$$
$$\langle hX, vY\rangle'_z = 0$$
$$\langle vX, vY\rangle'_z = \langle L_z(vX), L_z(vY)\rangle_{p_m(z)},$$

$X, Y \in T_z(TM)$, $z \in TM$.

Prove that $\langle\ ,\ \rangle'$ is an Hermitian metric with respect to F and then the corresponding Kaehler form $\Omega(X,Y) = \langle FX, Y\rangle'$, for any X, Y vector fields on TM, is symplectic (see Dombrowski (1962)).

(iii) Extend (i) and (ii) for $k = 2$ (see de León (1981)).

7. Let $(M, \langle\ ,\ \rangle)$ be a Riemannian manifold with Lèvi-Civita connection ∇. Then the tangent bundle T^2M of order 2 of M is isomorphic to the Whitney sum of TM with itself, that is, the mapping

$$S: T^2M \to TM \oplus TM$$
$$\tilde{\sigma}^2(0) \rightsquigarrow (\dot{\sigma}(0), (\nabla_{\dot{\sigma}}\dot{\sigma})(0))$$

is a diffeomorphism, and, consequently T^2M becomes via S a vector bundle over M. Write S in local coordinates and extend the result for an arbitrary k (see Dodson & Radivoiovici (1982) and de León & Vázquez Abal (1984)).

CHAPTER II

GENERALIZED CLASSICAL MECHANICS

II.1. - Introduction

The geometrical study of Classical Mechanics shows that the Hamiltonian (respectively, Lagrangian) formalism may be characterized by intrinsical structures canonically defined on the cotangent (respectively, tangent) bundle of a differentiable manifold. In terms of fibered manifolds these bundles may be viewed in the following form. If we consider Particle Mechanics as a Mechanics which involves only one independent variable (the time) then we distinguish:

(i) the autonomous (or time-independent) case, characterized by a fibered manifold (M,p,N) where $N = \mathbb{R}$ (the real numbers) and $M = \mathbb{R}\times E$, where E is a m-dimensional manifold. The velocity space corresponding to the configuration manifold E is the submanifold $J^1_o M$ of all 1-jets of mappings $s: \mathbb{R} \to M$ with source at the origin (that is, considering the local jet expression one has the coordinate $x_1 = 0$). Therefore, $J^1_o M = TE$, the tangent bundle of E. All Lagrangians are taken as real functions defined on TE and, locally, the coordinates are expressed by $(y^A, z^A_{(1)})$ (or $(q^A, \dot{q}^A)$). Along 1-jet prolongations we have $z^A_{(1)} = \frac{dq^A}{dt} = \dot{q}^A$.

(ii) the <u>non-autonomous</u> (or <u>time-dependent</u>) case, which corresponds to the situation where $x_1 = t$ is not necessarily zero. We have $J^1M = \mathbb{R}\times TE$.

Let us suppose that N is an arbitrary finite dimensional manifold and $E = \mathbb{R}$, with $M = N\times\mathbb{R}$. Then the submanifold for which the coordinates y^1 vanish is the cotangent bundle T^*N of N (the non-autonomous situation is developed on $\mathbb{R}\times T^*N$). Therefore, if we consider the fibrations $M = \mathbb{R}\times N \to \mathbb{R}$, $M = \mathbb{R}\times N \to N$, we have, respectively, $J^1M = \mathbb{R}\times TN$ (or TN, for the autonomous case) and $J^1M = R\times T^*N$ (or T^*N, for the autonomous case) and we may develop the usual description of Classical Mechanics on J^1M (Lagrangian and Hamiltonian formalism) if it is previously defined what fibration is considered.

A more general situation is obtained when many independent variables are considered. A <u>Classical Field Theory</u> is a "Mechanical" system involving many "times" (we recognize that such definition is not adequate in physical terms). The variables y^A are now functions of the set $(x_1,\ldots,x_n)$ and are called <u>field components</u> (or <u>variables</u>). We may have systems involving implicitly or explicitly the "times" $x_1,\ldots,x_n$.

The Lagrangian formalism is described by functions defined along J^1M, where now $M = N\times E$ is fibered over N and the Hamiltonian formalism is described along J^1M, with $M = N\times E$ fibered over E. Coordinate systems are, respectively, given by (x_a, y^A, z_a^A) and (y^A, x_a, t_A^a) with $t_A^a(\tilde{s}^1) = \partial(x_a\circ s)/\partial y^A$ and $t_A^a\circ z_a^A = \mathrm{Id}$ along jet prolongations. We represent such manifolds by J^1M and $J^{1*}M$, respectively.

Let us now recall some facts about Lagrangian and Hamiltonian formalisms (for more details, see Abraham & Marsden (1978), Arnold (1974) and Godbillon (1969)).

Concerning the Hamiltonian formalism, it is known that the theory is formalized under a special structure, called symplectic. Generically, a (regular) Hamiltonian system is defined by a triple (W,ω,H) where W is some given differentiable manifold of even dimension $2n$, ω is a closed two-form of maximal rank and H is a smooth function on W. The form ω is called symplectic and H Hamiltonian. The existence of such ω on W generates an isomorphism $s_\omega : \mathfrak{X}(W) \to$ $\to \Lambda^1(W)$ between the modules $\mathfrak{X}(W)$ and $\Lambda^1(W)$ of all vector fields and one forms on W, and a vector bundle isomorphism $s_\omega : TW \to T^*W$ between the tangent and cotangent bundles of W. In fact, we can define

$$s_\omega(X) = i_X\omega, \tag{1}$$

for any vector field X on W, that is,

$$s_\omega(X)(Y) = (i_X\omega)\ (Y) = \omega(X,Y),$$

for any vector fields X and Y on W.

Considering now the 1-form dH, the symplectic equation

$$i_X\omega = dH \tag{2}$$

has a unique solution X_H. If we take canonical coordinates (q^A, p_A) in W, that is, $\omega = dq^A \wedge dp_A$, one easily obtain, from (2), that the local expression of X_H is

$$X_H = \frac{\partial H}{\partial p_A}\frac{\partial}{\partial q^A} - \frac{\partial H}{\partial q^A}\frac{\partial}{\partial p_A}.$$

Consequently, the integral curves (trajectories) of X_H verify the well known <u>canonical equations</u>

$$\frac{\partial H}{\partial q^A} = - \frac{dp_A}{dt} , \quad \frac{\partial H}{\partial p_A} = \frac{dq^A}{dt} . \tag{3}$$

A particular (and fundamental) situation for Symplectic Mechanics concerns the cotangent bundle T^*M of a given manifold M. In such a case T^*M carries naturally a symplectic form $\omega_o = -d\lambda$, λ being the Liouville form on T^*M, defined in §3.6, Chap. I. Then ω_o is locally given by $\omega_o = dq^A \wedge \wedge dp_A$, (q^A,p_A) being the local coordinates in T^*M induced from the local coordinates (q^A) in M.

From the isomorphism (1) one defines the Poisson brackets: if f and g are functions on W then from the 1-forms df and dg we obtain the vector fields X_f and X_g, given by $i_{X_f}\omega = df$ and $i_{X_g}\omega = dg$, respectively, and we define

$$\{f,g\} = \omega(X_f,X_g). \tag{4}$$

Obviously, we have

$$\{f,g\} = -X_f(g) = X_g(f),$$

since ω is skew-symmetric.

In canonical coordinates (q^A,p_A), we have

$$\{f,g\} = \frac{\partial f}{\partial q^A}\frac{\partial g}{\partial p_A} - \frac{\partial f}{\partial p_A}\frac{\partial g}{\partial q^A} .$$

Concerning the Lagrangian formalism we may adopt two points of view. Since there is no symplectic form canonically defined on the tangent bundle TM of a given manifold M, we may transport to TM the symplectic form of T^*M via the

Legendre transformation Leg when the Lagrangian L is regular (that is, the Hessian of L with respect to the velocities is of maximal rank). We do this with the aid of the <u>fiber derivative</u> of L which is defined as follows. Denote by L_x the restriction of L to T_xM, $x \in M$. Then we can define a mapping $\text{Leg}: TM \to T^*M$ by

$$\text{Leg}(v) = dL_x(v), \qquad v \in T_xM,$$

where $dL_x(v): T_xM \to \mathbb{R}$ is the differential of L_x at v and T_xM and $T_v(T_xM)$ are canonically identified. In local coordinates, Leg is expressed by

$$\text{Leg}(q^A, \dot{q}^A) = (q^A, \frac{\partial L}{\partial \dot{q}^A}) \tag{5}$$

and will be called the fiber derivative of L. From (5), we easily deduce that Leg is a local diffeomorphism if and only if L is regular, that is, the Hessian matrix $(\frac{\partial^2 L}{\partial \dot{q}^A \partial \dot{q}^B})$ is invertible.

If we put

$$\omega_L = \text{Leg}^* \omega_o , \tag{6}$$

then ω_L is a closed 2-form on TM which will be symplectic if L is regular (see Abraham & Marsden (1978)).

An alternative was proposed by J. Klein in 1962. We may develop an intrinsical formalism <u>independent</u> of the regularity of L, with the aid of the almost tangent structure J canonically defined on TM. In fact, from (25), §3.5, Chap. I, (5) and (6), one easily deduces that the 2-form ω_L is now given by

$$\omega_L = -dd_J L .$$

Almost tangent geometry possesses numerous technical advantages over the more standard methods relying upon the fiber derivative and provides a natural framework for further generalizations which are studied in this chapter.

In Section 2, we will develop a generalized formalism for higher order Lagrangians. First, we recall on §2.1 the Tulczyjew's formalism; the Euler-Lagrange equations are obtained by means of a differential operator called the Lagrange differential, introduced by Tulczyjew (1975)(a). We clarify Tulczyjew's approach in the sense that the Lagrange differential can be directly defined in terms of the almost tangent structure of higher order.

The rest of the section is devoted to develop a similar study for higher order Lagrangians to the one elaborated by Klein. Again, the exterior calculus induced by the almost tangent structure of higher order permits us to construct a closed two-form on an appropriate jet bundle. In fact, if we begin with a Lagrangian L of order k, on a manifold M, we obtain a two-form ω_L on $T^{2k-1}M$, the tangent bundle of order $2k-1$ of M, which may be symplectic or, merely, presymplectic (notice that if M has dimension m, then $T^{2k-1}M$ has even dimension $2km$). In the later case, the canonical structures on higher tangent bundles are not sufficient to obtain the motion equations. In fact, we shall see that for this type of Mechanics a horizontal presymplectic structure subordinated to some hypothesis is necessary to the development of the theory. We proceed in such a way that, if $k = 1$, the Klein's formalism is reobtained.

The first attempts of developping this theory were proposed by Rodrigues (1975), (1976) and de León (1982)(a), (1982)(b), but the results were not satisfactory at all, since the choice of the (pre) symplectic form ω_L was inadequate (actually, we can see that this form appears as a particular case).

Section 3 is devoted to the development of the Hamiltonian form of the theory. The differential calculus on high tangent bundles permits to define the Legendre transformation and introduce the Jacobi-Ostrogradsky momentum coordinates for regular Lagrangians of higher order. From this, the corresponding canonical equations are derived in §3.2. We also recall the variational approach in §3.3; for a Lagrangian L of order k, the variational problem of higher order is equivalent to the variational problem of zero order. This has been proved by Dedecker (1979) under the hypothesis that L verifies the Helmholtz-Cartan regular condition. In the autonomous case, this is equivalent to the regularity of L.

To end the chapter, we define the Poisson brackets and study under what conditions a mapping is a canonical transformation. The Hamilton-Jacobi equation for this type of Mechanics is established.

II.2. - Generalized Lagrangian formalisms

2.1. Tulczyjew's formalism

Our aim in this section is to construct a differential operator (called the Lagrange differential) which will be used in order to obtain a Lagrangian formulation of Mechanics of higher order. To do this, we proceed as follows.

Let T^kM be the tangent bundle of order k, $k \geq 0$, of a manifold M, and denote by $\Lambda^q(T^kM)$ the real vector space of q-forms on T^kM. Then

$$\Lambda(T^kM) = \bigoplus_{q\geq 0} \Lambda^q(T^kM)$$

is the exterior algebra of differential forms on T^kM. For each $k' \leq k$, the canonical projection

$$\rho^k_{k'}: T^kM \to T^{k'}M$$

is a submersion. Then the adjoint mapping

$$(\rho^k_{k'})^*: \Lambda(T^{k'}M) \to \Lambda(T^kM)$$

is a canonical injection. Actually, if $k'' \leq k' \leq k$, then we have

$$(\rho^k_{k'})^* \circ (\rho^{k'}_{k''})^* = (\rho^k_{k''})^*.$$

Let Λ denote the quotient set of $\bigcup_{k\geq 0} \Lambda(T^kM)$ by the equivalence relation according to which two forms $\mu \in \Lambda(T^kM)$ and $\nu \in \Lambda(T^{k'}M)$ are equivalent if $k' \leq k$ and $\mu = (\rho^k_{k'})^*\nu$, or $k' \geq k$ and $\nu = (\rho^{k'}_k)^*\mu$. The exterior differential d and the exterior product $\wedge$ extend in a natural way to Λ in such a way that Λ becomes a commutative graded algebra. Consequently, derivations and skew-derivations on Λ can be considered.

Next, we introduce a derivation d_T on Λ due to Tulczyjew (1975)(a). First, for every function $f: T^kM \to \mathbb{R}$, we shall define a function $d_Tf: T^{k+1}M \to \mathbb{R}$ as follows:

$$(d_T f)(\tilde{\sigma}^{k+1}(0)) = df(\tilde{\sigma}^k(0))(j_k(\tilde{\sigma}^{k+1}(0))),$$

where $j_k\colon T^{k+1}M \to T(T^kM)$ is the canonical injection.

From (13), §3.2, chap. I, one deduces that $d_T f$ is locally given by

$$\begin{aligned}(d_T f)(z^A_{(0)}, z^A_{(1)}, \ldots, z^A_{(k+1)}) &= \sum_{i=0}^{k} (i+1) \frac{\partial f}{\partial z^A_{(i)}} z^A_{(i+1)} \\ &= \frac{\partial f}{\partial z^A_{(0)}} z^A_{(1)} + 2 \frac{\partial f}{\partial z^A_{(1)}} z^A_{(2)} + \ldots + (k+1) \frac{\partial f}{\partial z^A_{(k)}} z^A_{(k+1)}\end{aligned} \qquad (1)$$

Notice that the partial derivatives on the right hand of (1) are valued at the point $(z^A_{(0)}, z^A_{(1)}, \ldots, z^A_{(k)})$.

Now, one easily verifies that $d_T\colon f \rightsquigarrow d_T f$ acts as a derivation on the functions of Λ. Then, we have

PROPOSITION (1). d_T <u>extends to a derivation</u> d_T <u>on</u> Λ <u>of degree</u> 0 (<u>or to a derivation on</u> Λ <u>of type</u> d_* <u>and degree</u> 0 <u>in the sense of Frölicher & Nijenhuis</u> (1956)).

Since $d_T d = d d_T$, the derivation d_T is completely determined by (1) and the following relations:

$$d_T dz^A_{(i)} = (i+1)\, dz^A_{(i+1)}, \qquad 0 \le i \le k. \qquad (2)$$

In the sequel, we shall extend the derivations and skew-derivations introduced in the Chapter I to Λ. In order to do this, we shall adopt the following convention. If $r > k$, we put $J_r = 0$ on $T(T^kM)$; if $k = 0$, then T^oM is identified to M and we take $J_o = \mathrm{Id}$. Therefore, the diagram

$$\begin{array}{ccc} T(T^kM) & \xrightarrow{J_r} & T(T^kM) \\ \downarrow{\scriptstyle T\rho^k_{k'}} & & \downarrow{\scriptstyle T\rho^k_{k'}} \\ T(T^{k'}M) & \xrightarrow{J_r} & T(T^{k'}M) \end{array} \tag{3}$$

is commutative.

From the commutativity of (3), we deduce that the diagrams

$$\begin{array}{ccc} \Lambda(T^{k'}M) & \xrightarrow{i_{J_r}} & \Lambda(T^{k'}M) \\ \downarrow{\scriptstyle (\rho^k_{k'})^*} & & \downarrow{\scriptstyle (\rho^k_{k'})^*} \\ \Lambda(T^kM) & \xrightarrow{i_{J_r}} & \Lambda(T^kM) \end{array}$$

and

$$\begin{array}{ccc} \Lambda(T^{k'}M) & \xrightarrow{d_{J_r}} & \Lambda(T^{k'}M) \\ \downarrow{\scriptstyle (\rho^k_{k'})^*} & & \downarrow{\scriptstyle (\rho^k_{k'})^*} \\ \Lambda(T^kM) & \xrightarrow{d_{J_r}} & \Lambda(T^kM) \end{array}$$

are also commutative.

Hence, the mappings i_{J_r} (resp., d_{J_r}) extend to a derivation on Λ, i_{J_r} (resp., a skew-derivation on Λ, d_{J_r}) of degree 0 (resp., of degree 1).

As above, we now put $C_r = 0$, if $r > k$, on T^kM. Then, the diagrams

$$\begin{array}{ccc}
\Lambda(T^{k'}M) & \xrightarrow{\ i_{C_r}\ } & \Lambda(T^{k'}M) \\
\downarrow (\rho^{k}_{k'})^{*} & & \downarrow (\rho^{k}_{k'})^{*} \\
\Lambda(T^{k}M) & \xrightarrow{\ i_{C_r}\ } & \Lambda(T^{k}M)
\end{array}$$

and

$$\begin{array}{ccc}
\Lambda(T^{k'}M) & \xrightarrow{\ \mathcal{L}_{C_r}\ } & \Lambda(T^{k'}M) \\
\downarrow (\rho^{k}_{k'})^{*} & & \downarrow (\rho^{k}_{k'})^{*} \\
\Lambda(T^{k}M) & \xrightarrow{\ \mathcal{L}_{C_r}\ } & \Lambda(T^{k}M)
\end{array}$$

are commutative.

Hence, the mappings i_{C_r} (resp., $\mathcal{L}_{C_r}$) extend to a skew-derivation on Λ, i_{C_r} (resp., a derivation on Λ, $\mathcal{L}_{C_r}$) of degree -1 (resp., 0).

REMARK (1). If we adopt the terminology of Frölicher and Nijenhuis, then i_{J_r} and i_{C_r} are derivations of type i_* and degree 0 and -1, respectively, and d_{J_r} and $\mathcal{L}_{C_r}$ are derivations of type d_* and degree 1 and 0, respectively, on Λ.

Next, we shall complete the differential calculus developed in the Chapter I. So, we have

PROPOSITION (2). For each r, $r \geq 1$,

$$[i_{J_r}, d_T] = i_{J_r} d_T - d_T i_{J_r} = i_{J_{r-1}} .$$

<u>Proof</u>: The commutator $[i_{J_r}, d_T] = i_{J_r} d_T - d_T i_{J_r}$ is a derivation which acts trivially on the functions. Then the result follows from (24), §3.4, Chap. I, and (2). □

PROPOSITION (3). (i) <u>We have</u>

$$[\mathcal{L}_{C_1}, d_T] = \mathcal{L}_{C_1} d_T - d_T \mathcal{L}_{C_1} = d_T$$

(ii) <u>Moreover, for each</u> r, $r > 1$, <u>we have</u>

$$[\mathcal{L}_{C_r}, d_T] = \mathcal{L}_{C_r} d_T - d_T \mathcal{L}_{C_r} = (r-1)\mathcal{L}_{C_{r-1}} .$$

<u>Proof</u>: Indeed, for each coordinate function $z^A_{(i)}$, we obtain

$$\begin{aligned}
[\mathcal{L}_{C_1}, d_T] z^A_{(i)} &= \mathcal{L}_{C_1} d_T (z^A_{(i)}) - d_T \mathcal{L}_{C_1} (z^A_{(i)}) \\
&= (i+1)^2 \; z^A_{(i+1)} - i(i+1) \; z^A_{(i+1)} \\
&= (i+1) \; z^A_{(i+1)} \\
&= d_T (z^A_{(i)}),
\end{aligned}$$

and, for each function f,

$$\begin{aligned}
[\mathcal{L}_{C_1}, d_T] df &= \mathcal{L}_{C_1} d_T df - d_T \mathcal{L}_{C_1} df \\
&= d\mathcal{L}_{C_1} d_T f - dd_T \mathcal{L}_{C_1} f \\
&= d[\mathcal{L}_{C_1}, d_T] f \\
&= dd_T f \\
&= d_T df,
\end{aligned}$$

since $d_T d = dd_T$ and $\mathcal{L}_{C_1} d = d\mathcal{L}_{C_1}$.

Because $[\mathcal{L}_{C_1}, d_T]$ and d_T are both derivations, the result easily follows.

The proof of (ii) is similar and is left to the reader as an exercise. □

Next, we study the action of d_T on homogeneous forms.

PROPOSITION (4). <u>If</u> μ <u>is an homogeneous q-form on</u> $T^k M$ <u>of degree</u> α, <u>then</u> $d_T \mu$ <u>is an homogeneous q-form on</u> $T^{k+1} M$ <u>of</u>

degree $\alpha+1$.

Proof: In fact, from Proposition (3), we have

$$\begin{aligned}\mathcal{L}_{C_1} d_T\mu &= d_T\mu + d_T\mathcal{L}_{C_1}\mu \\ &= d_T\mu + \alpha\, d_T\mu \\ &= (\alpha+1)d_T\mu. \quad \square\end{aligned}$$

Now, we shall adopt the following definition.

DEFINITION (1). The operator

$$\delta = (i_{J_o} - d_T i_{J_1} + \frac{1}{2!} d_T^2 i_{J_2} - \frac{1}{3!} d_T^3 i_{J_3} + \ldots)d \tag{4}$$

will be called the Lagrange differential (Tulczyjew, (1975)).

REMARK (2). The right hand of (4) is meaningful because the number of terms which are not zero is finite. In fact, for a given form μ on T^kM, $i_{J_r}\mu = 0$, if $r > k$.

REMARK (3). J. Peetre defines another differential operator (which he calls the Euler derivative) and claims that this operator and the one used by Tulczyjew are the same. For more details, we remit to Peetre (1978).

Now, let L be a Lagrangian of order k on a given manifold M, that is, $L: T^kM \to \mathbb{R}$ is a function on the tangent bundle of order k of M. Then, we have

$$\delta L = (i_{J_o} - d_T i_{J_1} + \frac{1}{2!} d_T^2 i_{J_2} + \ldots + (-1)^k \frac{1}{k!} d_T^k i_{J_k})dL \tag{5}$$

and, hence, δL is a 1-form on $T^{2k}M$.

Notice that $i_{J_r} dL = d_{J_r} L$, because i_{J_r} acts trivially on the functions. Then (5) can be equivalently written as follows:

$$\delta L = dL - d_T d_{J_1} L + \frac{1}{2!} d_T^2 d_{J_2} L + \dots + (-1)^k \frac{1}{k!} d_T^k d_{J_k} L . \qquad (5)'$$

If we now suppose that σ is a curve in M with local coordinates $(q^A(t))$, then it is easy to prove that, along the prolongation $\tilde{\sigma}^{2k}$ of σ to $T^{2k}M$, (5) can be written as follows:

$$\delta L = \sum_{A=1}^{m} \left(\frac{\partial L}{\partial q^A} - \frac{d}{dt}\frac{\partial L}{\partial \dot{q}^A} + \frac{d^2}{dt^2}\frac{\partial L}{\partial \ddot{q}^A} + \dots + (-1)^k \frac{d^k}{dt^k}\frac{\partial L}{\partial q^{(k)A}}\right) dq^A , \qquad (6)$$

where $\dot{q}^A = \frac{dq^A}{dt}$, $\ddot{q}^A = \frac{d^2q^A}{dt^2}$, ..., $q^{(k)A} = \frac{d^kq^A}{dt^k}$, $1 \le A \le m$, $\dim M = m$.

For sake of simplicity, we only prove (6) in the case $k = 2$. The general case can be proved by the reader without major difficulties and it is left as an exercise. Then, if $k = 2$, we have

$$\delta L = (i_{J_o} - d_T i_{J_1} + \frac{1}{2} d_T^2 i_{J_2}) dL . \qquad (7)$$

Since

$$dL = \frac{\partial L}{\partial z^A_{(0)}} dz^A_{(0)} + \frac{\partial L}{\partial z^A_{(1)}} dz^A_{(1)} + \frac{\partial L}{\partial z^A_{(2)}} dz^A_{(2)} ,$$

we obtain

$$i_{J_o} dL = \frac{\partial L}{\partial z^A_{(0)}} dz^A_{(0)} + \frac{\partial L}{\partial z^A_{(1)}} dz^A_{(1)} + \frac{\partial L}{\partial z^A_{(2)}} dz^A_{(2)} ,$$

$$i_{J_1} dL = \frac{\partial L}{\partial z^A_{(1)}} dz^A_{(0)} + \frac{\partial L}{\partial z^A_{(2)}} dz^A_{(1)}$$

$$i_{J_2} dL = \frac{\partial L}{\partial z^A_{(2)}} dz^A_{(0)} ,$$

taking into account (24), §3.4, Chap. I. From (2), one easily deduces that

$$\delta L = \frac{\partial L}{\partial z^A_{(0)}} dz^A_{(0)} + \frac{\partial L}{\partial z^A_{(1)}} dz^A_{(1)} + \frac{\partial L}{\partial z^A_{(2)}} dz^A_{(2)}$$

$$- d_T\left(\frac{\partial L}{\partial z^A_{(1)}}\right) dz^A_{(0)} - \frac{\partial L}{\partial z^A_{(1)}} dz^A_{(1)} - d_T\left(\frac{\partial L}{\partial z^A_{(2)}}\right) dz^A_{(1)}$$

$$- 2\frac{\partial L}{\partial z^A_{(2)}} dz^A_{(2)} + \frac{1}{2} d_T^2 \left(\frac{\partial L}{\partial z^A_{(2)}}\right) dz^A_{(0)}$$

$$+ d_T\left(\frac{\partial L}{\partial z^A_{(2)}}\right) dz^A_{(1)} + \frac{\partial L}{\partial z^A_{(2)}} dz^A_{(2)}$$

$$= \left(\frac{\partial L}{\partial z^A_{(0)}} - d_T\left(\frac{\partial L}{\partial z^A_{(1)}}\right) + \frac{1}{2} d_T^2 \left(\frac{\partial L}{\partial z^A_{(2)}}\right)\right) dz^A_{(0)} .$$

Now, as $d_T\left(\frac{\partial L}{\partial z^A_{(1)}}\right) = \frac{d}{dt}\left(\frac{\partial L}{\partial \dot{q}^A}\right)$ and $\frac{1}{2} d_T^2\left(\frac{\partial L}{\partial z^A_{(2)}}\right) = \frac{d^2}{dt^2}\left(\frac{\partial L}{\partial \ddot{q}^A}\right)$ along $\tilde{\sigma}^4$, one finally obtains

$$\delta L = \sum_{A=1}^{m} \left(\frac{\partial L}{\partial q^A} - \frac{d}{dt}\frac{\partial L}{\partial \dot{q}^A} + \frac{d^2}{dt^2}\frac{\partial L}{\partial \ddot{q}^A}\right) dq^A ,$$

along $\tilde{\sigma}^4$.

Notice that $\delta L = 0$, along $\tilde{\sigma}^4$, if and only if

$$\frac{\partial L}{\partial q^A} - \frac{d}{dt}\frac{\partial L}{\partial \dot{q}^A} + \frac{d^2}{dt^2}\frac{\partial L}{\partial \ddot{q}^A} = 0, \qquad 1 \le A \le m,$$

that is, σ satisfies the Euler-Lagrange equations.

Let us now return to the general case. Then we have

THEOREM (1). Let L be a Lagrangian of order k and σ a curve in M. Then, $\delta L = 0$, along $\tilde{\sigma}^{2k}$, if and only if σ satisfies the Euler-Lagrange equations

$$\frac{\partial L}{\partial q^A} - \frac{d}{dt}\frac{\partial L}{\partial \dot{q}^A} + \frac{d^2}{dt^2}\frac{\partial L}{\partial \ddot{q}^A} + \ldots + (-1)^k \frac{d^k}{dt^k}\frac{\partial L}{\partial q^{(k)A}} = 0, \qquad (8)$$

$1 \le A \le m$.

2.2. Horizontal presymplectic structures

In this section, we shall consider a certain kind of geometric structure which includes the symplectic structure as a particular case and will be used in order to obtain the motion equations for Lagrangians depending on higher derivatives.

Now, suppose that W is a manifold of dimension $2n+r$, $n \geq 1$ and $r \geq 0$, endowed with a 2-form ω of constant rank $2n$; ω is said to be an almost presymplectic form and the pair (W,ω) an almost presymplectic manifold. If, in addition, ω is closed, ω is called a presymplectic form and (W,ω) a presymplectic manifold. Clearly, when $r = 0$, (W,ω) is a symplectic manifold.

REMARK (4). An important application of presymplectic geometry to the theory of constrained classical systems can be found in Gotay (1979), Gotay & Nester (1979), Pnevmatikos (1983) and Rodrigues (1984).

For an almost presymplectic manifold (W,ω) we can also consider a linear mapping $s_\omega: \mathfrak{X}(W) \to \Lambda^1(W)$ and a vector bundle homomorphism $s_\omega: TW \to T^*W$, defined by (1), §2.1, but the s_ω are not necessarily isomorphisms. Now, suppose that $P: \mathfrak{X}(W) \to \mathfrak{X}(W)$ is a projection operator on $\mathfrak{X}(W)$, that is, $P^2 = P$ (we shall denote by the same symbol the corresponding projection operator $P: TW \to TW$ on TW). We put $Q = Id-P$. Then we know that

$$\mathfrak{X}(W) = \mathfrak{X}_P(W) \oplus \mathfrak{X}_Q(W),$$

where $\mathfrak{X}_P(W) = \text{Im } P$ and $\mathfrak{X}_Q(W) = \text{Im } Q$, or, if we want,

$\mathfrak{X}_P(W) = \text{Ker } Q$ and $\mathfrak{X}_Q(W) = \text{Ker } P$. (Obviously, we also have

$$TW = T_P W \oplus T_Q W$$

where $T_P W = P(TW)$ and $T_Q W = Q(TW)$). We also have that

$$\Lambda^1(W) = \Lambda^1_P(W) \oplus \Lambda^1_Q(W),$$

where $\Lambda^1_P(W)$ and $\Lambda^1_Q(W)$ are defined in a similar way from the adjoint mappings $P^*(\alpha) = \alpha \circ P$ and $Q^*(\alpha) = \alpha \circ Q$. (Once more, we have

$$T^*W = T^*_P W \oplus T^*_Q W,$$

where $T^*_P W$ and $T^*_Q W$ are the obvious ones).

DEFINITION (2). Let W, ω and P as above. We say that (ω, P) is an <u>almost horizontal presymplectic structure</u> on W if

(i) $\dim \mathfrak{X}_Q(W) = r$

(ii) $\text{Ker } s_\omega = \text{Ker } P = \mathfrak{X}_Q(W)$.

If, in addition, ω is closed, then (ω, P) is said to be a <u>horizontal presymplectic structure</u> on W.

REMARK (5). A geometric structure defined by a projection operator P on W is called an almost product structure on W. (See Walker (1961), Willmore (1956), Lehman-Lejeune (1964), etc). Consequently, a horizontal presymplectic structure (ω, P) on W is a presymplectic structure ω plus an almost product structure on W satisfying the conditions (i) and (ii) of Definition (2) (for more details on this kind of geometric structures, see de Barros (1964), (1965), (1967)).

REMARK (6). If we suppose that all manifolds under consideration are paracompact, we have a Riemannian metric $\langle\ ,\ \rangle$ on W. Therefore, if ω is a presymplectic form on W, we can take $T_P W$ as the orthogonal complement of $\operatorname{Ker} s_\omega$ with respect to $\langle\ ,\ \rangle$ in each point of W and so we have a horizontal presymplectic structure (ω,P) on W.

PROPOSITION (5). Let (ω,P) be an almost horizontal presymplectic structure on W and $f\colon W \to \mathbb{R}$ be a function. Then there is only one vector field $X \in \operatorname{Im} P$ such that

$$i_X\omega = P^*(df), \tag{9}$$

where P^* is the adjoint of P. In particular if $df \in \operatorname{Im} P^*$, then (9) takes the form (2), §2, that is,

$$i_X\omega = df. \tag{9)'$$

Proof: Let us take the linear mapping

$$s_\omega\colon \mathfrak{X}_P(W) \oplus \mathfrak{X}_Q(W) \to \Lambda^1_P(W) \oplus \Lambda^1_Q(W)$$
$$X \rightsquigarrow i_X\omega$$

Taking into account that $\operatorname{Ker} s_\omega = \mathfrak{X}_Q(W)$, it follows that the restriction of s_ω to the subspace $\mathfrak{X}_P(W)$ is one to one and so $s_\omega(\mathfrak{X}_P(W)) = \Lambda^1_P(W)$, since we have $s_\omega(X) = P^*s_\omega(X)$, for any vector field $X = PX$. In fact, since $\operatorname{Ker} s_\omega = \mathfrak{X}_Q(W)$ and ω is skew-symmetric, we have

$$\begin{aligned} s_\omega(X)(Y) &= s_\omega(X)(PY+QY) \\ &= s_\omega(X)(PY) + s_\omega(X)(QY) \\ &= s_\omega(X)(PY) \\ &= (P^*s_\omega(X))(Y), \end{aligned}$$

for any vector field Y on W. This shows that $s_\omega(X) =$

$= P^*s_\omega(X)$ and so s_ω induces an isomorphism $\hat{s}_\omega$ between $\mathfrak{X}_P(W)$ and $\Lambda^1_P(W)$ (and, obviously, between T_PW and T^*_PW) which gives (9). The equality (9)' comes from the fact that $df = P^*(df) + Q^*(df)$ and so $Q^*(df) = 0$ where $df \in \mathrm{Im}\, P^*$. □

REMARK (7). It is clear now that we can also define the Poisson brackets for the present situation. We put

$$X_f = (\hat{s}_\omega)^{-1}(P^*df),$$

and, then, use (4), §1.

REMARK (8). If we consider a Hamiltonian $H: W \to \mathbb{R}$, then (9) corresponds to the symplectic expression of Hamilton equations of motion for a constraint manifold in the sense of Dirac mechanics.

2.3. Generalized Klein's formalism

In this section, we shall develop a Lagrangian formalism for the Mechanics of higher order which generalizes Klein's Lagrangian formalism for the Classical Mechanics.

Now, let L be a Lagrangian of order k on a given manifold M, that is, $L: T^kM \to R$ is a function on T^kM. We put

$$\omega_L = -dd_{J_1}L + \frac{1}{2!}\, d_T dd_{J_2}L - \frac{1}{3!}\, d_T^2 dd_{J_3}L + \dots + (-1)^k \frac{1}{k!}\, d_T^{k-1}\, dd_{J_k}L\,. \tag{10}$$

Then ω_L is a closed 2-form on $T^{2k-1}M$, (a reason for the above definition will be presented in §4.1).

Next, the canonical vector fields on higher tangent bundles permit us to define the <u>energy</u> E of L as follows:

$$E = C_1L - \frac{1}{2!}\, d_T(C_2L) + \frac{1}{3!}\, d_T^2(C_3L)$$
$$+\dots+ (-1)^{k-1}\frac{1}{k!}\, d_T^{k-1}(C_kL) - L\,. \qquad (11)$$

So, $E\colon T^{2k-1}M \to \mathbb{R}$ is a function on $T^{2k-1}M$.

REMARK (9). We can easily check that the choice of E is appropriate. In fact, take $k = 2$, for instance. Then, if σ is a curve in M, E is expressed along the prolongation $\tilde{\sigma}^3$ of σ to T^3M as follows:

$$E = \left(\frac{\partial L}{\partial \dot{q}^A} - \frac{d}{dt}\frac{\partial L}{\partial \ddot{q}^A}\right)\dot{q}^A + \frac{\partial L}{\partial \ddot{q}^A}\,\ddot{q}^A - L$$
$$= p_{A/1}\,\dot{q}^A + p_{A/2}\,\ddot{q}^A - L,$$

where $(q^A(t))$ are the local coordinates of σ and $p_{A/1}$, $p_{A/2}$ are the Jacobi-Ostrogradsky momentum coordinates defined in §4.1.

For an arbitrary k, we obtain

$$E = \sum_{i=1}^{k} p_{A/i}\, q^{(i)A} - L,$$

along $\tilde{\sigma}^{2k-1}$, where $p_{A/1},\dots,p_{A/k}$ are the corresponding Jacobi-Ostrogradsky momentum coordinates (see §4.1). Therefore, E is, in fact, the Hamiltonian energy corresponding to L (see §4.2).

Return now to the Lagrangian L. In the sequel, we will analize the equation

$$i_\xi \omega_L = dE\,. \qquad (12)$$

(We will see that (12) is, in fact, the intrinsical form of Lagrangian equations of motion).

Let us recall that $T^{2k-1}M$ has even dimension $2km$, $\dim M = m$, and, hence, ω_L may be symplectic. In this case, the Lagrangian L is said to be <u>regular</u>; otherwise, L is <u>degenerate</u>, <u>irregular</u> or even, <u>singular</u>. This terminology used here can be justified as follows. Indeed, in the case $k = 1$, ω_L becomes simply

$$\omega_L = \frac{\partial^2}{\partial \dot{q}^A \partial q^B} dq^A \wedge dq^B + \frac{\partial^2 L}{\partial \dot{q}^A \partial \dot{q}^B} dq^A \wedge d\dot{q}^B ,$$

in local coordinates $(q^A, \dot{q}^A)$, and, so,

$$\omega_L^m = \overbrace{\omega_L \wedge \ldots \wedge \omega_L}^{m}$$

$$= c \det \left(\frac{\partial^2 L}{\partial \dot{q}^A \partial \dot{q}^B}\right) dq^1 \wedge \ldots \wedge dq^m \wedge d\dot{q}^1 \wedge \ldots \wedge d\dot{q}^m, \qquad (13)$$

where c is some non-zero constant. Hence, (13) shows that ω_L <u>is symplectic if and only if the velocity Hessian</u> $\left(\frac{\partial^2 L}{\partial \dot{q}^A \partial \dot{q}^B}\right)$ <u>of</u> L <u>is invertible</u> (that is, of maximal rank).

Now, take $k = 2$. A long but straightforward calculation leads to

$$\begin{aligned}
\omega_L = {} & \left(\frac{\partial^2 L}{\partial z^A_{(1)} \partial z^B_{(0)}} - \frac{1}{2} d_T\left(\frac{\partial^2 L}{\partial z^A_{(2)} \partial z^B_{(0)}}\right)\right) dz^A_{(0)} \wedge dz^B_{(0)} \\
& + \left(\frac{\partial^2 L}{\partial z^A_{(1)} \partial z^B_{(1)}} - \frac{1}{2} \frac{\partial^2 L}{\partial z^A_{(0)} \partial z^B_{(2)}} - \frac{1}{2} \frac{\partial^2 L}{\partial z^A_{(2)} \partial z^B_{(0)}} \right. \\
& \qquad \left. - \frac{1}{2} d_T\left(\frac{\partial^2 L}{\partial z^A_{(2)} \partial z^B_{(1)}}\right)\right) dz^A_{(0)} \wedge dz^B_{(1)} \\
& + \left(\frac{\partial^2 L}{\partial z^A_{(1)} \partial z^B_{(2)}} - \frac{\partial^2 L}{\partial z^A_{(2)} \partial z^B_{(1)}} - \frac{1}{2} d_T\left(\frac{\partial^2 L}{\partial z^A_{(2)} \partial z^B_{(2)}}\right)\right) dz^A_{(0)} \wedge dz^B_{(2)} \\
& + \left(\frac{\partial^2 L}{\partial z^A_{(2)} \partial z^B_{(1)}} - \frac{1}{2} \frac{\partial^2 L}{\partial z^A_{(1)} \partial z^B_{(1)}}\right) dz^A_{(1)} \wedge dz^B_{(1)}
\end{aligned}$$

$$+ \frac{1}{2} \frac{\partial^2 L}{\partial z^A_{(2)} \partial z^B_{(2)}} dz^A_{(1)} \wedge dz^B_{(2)}$$

$$- \frac{3}{2} \frac{\partial^2 L}{\partial z^A_{(2)} \partial z^B_{(2)}} dz^A_{(0)} \wedge dz^B_{(3)} . \qquad (14)$$

From (14), we deduce that

$$\omega_L^{2m} = \overbrace{\omega_L \wedge \ldots \wedge \omega_L}^{2m} = c' \det\left(\frac{\partial^2 L}{\partial z^A_{(2)} \partial z^B_{(2)}}\right) dz^1_{(0)} \wedge \ldots \wedge dz^m_{(3)},$$

where c' is a non-zero constant, and, consequently, ω_L is symplectic if and only if the matrix $\left(\frac{\partial^2 L}{\partial z^A_{(2)} \partial z^B_{(2)}}\right)$ is invertible.

For an arbitrary k, the same result remains valid and, from a tedious direct calculation, we can see that ω_L <u>is symplectic if and only if the matrix</u> $\left(\frac{\partial^2 L}{\partial z^A_{(k)} \partial z^B_{(k)}}\right)$ <u>is invertible</u>.

Next, we shall discuss separately the cases of regular or irregular Lagrangians.

(a). - <u>The Lagrangian</u> L <u>is regular</u>

In this case, ω_L is symplectic and then the equation (12) possesses a unique solution ξ. We shall see that the integral curves of ξ are the dynamical trajectories corresponding to the Lagrangian L (see the Theorems (2) and (3)).

For $k = 1$, we know that the unique solution ξ of (12) is automatically a semispray (or a second order equation). In fact, if $k = 1$, (10) and (11) reduce to $\omega_L = -dd_J L$ and $E = CL - L$, respectively. Then we have

$$(i_\xi \omega_L)(X) = \omega_L(\xi, X)$$
$$= -dd_J L(\xi, X)$$
$$= -\xi(JX)L + X(J_1\xi)L + J_1(\xi, X)L$$

and

$$dE(X) = XE = X(CL) - XL,$$

for any vector field X on TM.

Let us suppose that ξ is locally expressed by

$$\xi = \xi_1^B \frac{\partial}{\partial q^B} + \xi_2^B \frac{\partial}{\partial \dot{q}^B}.$$

If we take $X = \frac{\partial}{\partial \dot{q}^A}$, one easily deduces

$$(i_\xi \omega_L)\left(\frac{\partial}{\partial \dot{q}^A}\right) = \xi_1^B \frac{\partial^2 L}{\partial \dot{q}^B \partial \dot{q}^A} \quad \text{and} \quad dE\left(\frac{\partial}{\partial \dot{q}^A}\right) = \dot{q}^B \frac{\partial^2 L}{\partial \dot{q}^B \partial \dot{q}^A}.$$

From (12), we get $\xi_1^B = \dot{q}^B$ because the velocity Hessian $\left(\frac{\partial^2 L}{\partial \dot{q}^A \partial \dot{q}^B}\right)$ is invertible (Godbillon (1969)), and then ξ is a semispray on TM.

Next, we extend this result for regular Lagrangians of higher order.

PROPOSITION (6). Let L be a regular Lagrangian of order k on M. Then the vector field ξ given by (12) is a semispray on $T^{2k-1}M$ of type 1, that is, $J_1\xi = C_1$.

Proof: For sake of simplicity, we only prove the proposition in the case $k = 2$. For an arbitrary $k > 2$, it easily follows by induction on the type of the semispray. Now, take $k = 2$. Then ω_L and E, given by (10) and (11), respectively, reduce to $\omega_L = -dd_{J_1}L + \frac{1}{2} dd_T d_{J_2} L$ and $E = C_1 L - \frac{1}{2} d_T(C_2 L) - L$, respectively. Then, whe have

$$(i_\xi \omega_L)(X) = \omega_L(\xi, X)$$
$$= -dd_{J_1} L(\xi, X) + \frac{1}{2} dd_T d_{J_2} L(\xi, X)$$
$$= -\xi(J_1 X)L + X(J_1 \xi)L + J_1[\xi, X]L$$
$$+ \frac{1}{2} \xi(d_T d_{J_2} L(X)) - \frac{1}{2} X(d_T d_{J_2} L(\xi))$$
$$- \frac{1}{2} d_T d_{J_2} L[\xi, X] \tag{15}$$

and

$$dE(X) = XE = X(C_1 L) - \frac{1}{2} X\, d_T(C_2 L) - XL, \tag{16}$$

for any vector field X on $T^{2k-1}M$.

We make the proof in three steps.

<u>Step 1</u>: Take $X = \frac{\partial}{\partial z^A_{(3)}}$. If we suppose that the vector field ξ given by (12) is locally expressed by

$$\xi = \xi^B_{(1)} \frac{\partial}{\partial z^B_{(0)}} + \xi^B_{(2)} \frac{\partial}{\partial z^B_{(1)}} + \xi^B_{(3)} \frac{\partial}{\partial z^B_{(2)}} + \xi^B_{(4)} \frac{\partial}{\partial z^B_{(3)}},$$

where $\xi^B_{(\ell)} = \xi^B_{(\ell)}(z^C_{(0)}, z^C_{(1)}, z^C_{(2)}, z^C_{(3)})$, then we have

$$[\xi, \frac{\partial}{\partial z^A_{(3)}}] = \frac{\partial \xi^B_{(1)}}{\partial z^A_{(3)}} \frac{\partial}{\partial z^B_{(0)}} - \frac{\partial \xi^B_{(2)}}{\partial z^A_{(3)}} \frac{\partial}{\partial z^B_{(1)}} -$$
$$- \frac{\partial \xi^B_{(3)}}{\partial z^A_{(3)}} \frac{\partial}{\partial z^B_{(2)}} - \frac{\partial \xi^B_{(4)}}{\partial z^A_{(3)}} \frac{\partial}{\partial z^B_{(3)}}$$

and

$$J_1[\xi, \frac{\partial}{\partial z^A_{(3)}}] = - \frac{\partial \xi^B_{(1)}}{\partial z^A_{(3)}} \frac{\partial}{\partial z^B_{(1)}} - \frac{\partial \xi^B_{(2)}}{\partial z^A_{(3)}} \frac{\partial}{\partial z^B_{(2)}}$$
$$- \frac{\partial \xi^B_{(3)}}{\partial z^A_{(3)}} \frac{\partial}{\partial z^B_{(3)}}.$$

Now, since $d_T d_{J_2} L$ is locally given by

$$d_T d_{J_2} L = d_T\left(\frac{\partial L}{\partial z^C_{(2)}}\right) dz^C_{(0)} + \frac{\partial L}{\partial z^C_{(2)}} dz^C_{(1)} ,$$

we obtain

$$d_T d_{J_2} L\left(\frac{\partial}{\partial z^A_{(3)}}\right) = 0,$$

$$d_T d_{J_2} L(\xi) = \xi^B_{(1)}\, d_T\left(\frac{\partial L}{\partial z^B_{(2)}}\right) + \xi^B_{(2)} \frac{\partial L}{\partial z^B_{(2)}}$$

and

$$d_T d_{J_2} L\left[\xi , \frac{\partial}{\partial z^A_{(3)}}\right] = - \frac{\partial \xi^B_{(1)}}{\partial z^A_{(3)}}\, d_T\left(\frac{\partial L}{\partial z^B_{(2)}}\right) - \frac{\partial \xi^B_{(2)}}{\partial z^A_{(3)}} \frac{\partial L}{\partial z^B_{(2)}} .$$

Consequently, if $X = \frac{\partial}{\partial z^A_{(3)}}$, (15) becomes

$$\frac{\partial}{\partial z^A_{(3)}}\left(\xi^B_{(1)} \frac{\partial L}{\partial z^B_{(1)}} + \xi^B_{(2)} \frac{\partial L}{\partial z^B_{(2)}}\right) - \frac{\partial \xi^B_{(1)}}{\partial z^A_{(3)}} \frac{\partial L}{\partial z^B_{(1)}} - \frac{\partial \xi^B_{(2)}}{\partial z^A_{(3)}} \frac{\partial L}{\partial z^B_{(2)}}$$

$$- \frac{1}{2} \frac{\partial}{\partial z^A_{(3)}}\left(\xi^B_{(1)}\, d_T\left(\frac{\partial L}{\partial z^B_{(2)}}\right)\right) - \frac{1}{2} \frac{\partial}{\partial z^A_{(3)}}\left(\xi^B_{(2)} \frac{\partial L}{\partial z^B_{(2)}}\right)$$

$$+ \frac{1}{2} \frac{\partial \xi^B_{(1)}}{\partial z^A_{(3)}}\, d_T\left(\frac{\partial L}{\partial z^B_{(2)}}\right) + \frac{1}{2} \frac{\partial \xi^B_{(2)}}{\partial z^A_{(3)}} \frac{\partial L}{\partial z^B_{(2)}}$$

$$= - \frac{1}{2} \xi^B_{(1)} \frac{\partial}{\partial z^A_{(3)}}\left(d_T\left(\frac{\partial L}{\partial z^B_{(2)}}\right)\right)$$

$$= - \frac{3}{2} \xi^B_{(1)} \frac{\partial^2 L}{\partial z^B_{(2)} \partial z^A_{(2)}} ,$$

since $L = L(z^C_{(0)}, z^C_{(1)}, z^C_{(2)})$.

On the other hand, if $X = \frac{\partial}{\partial z^A_{(3)}}$, (16) becomes

$$dE\left(\frac{\partial}{\partial z^A_{(3)}}\right) = - \frac{3}{2}\, z^B_{(1)} \frac{\partial^2 L}{\partial z^B_{(2)} \partial z^A_{(2)}} .$$

From (12), and taking into account the regurality of L, we deduce that $\xi^B_{(1)} = z^B_{(1)}$, $1 \le B \le m$, and, hence, ξ is a semisprazy of type 3.

<u>Step 2</u>: Take $X = \dfrac{\partial}{\partial z^A_{(2)}}$. Since $\xi = z^B_{(1)} \dfrac{\partial}{\partial z^B_{(0)}} + \xi^B_{(2)} \dfrac{\partial}{\partial z^B_{(1)}} + \xi^B_{(3)} \dfrac{\partial}{\partial z^B_{(2)}} + \xi^B_{(4)} \dfrac{\partial}{\partial z^B_{(3)}}$, one has

$$[\xi, \frac{\partial}{\partial z^A_{(2)}}] = - \frac{\partial \xi^B_{(2)}}{\partial z^A_{(2)}} \frac{\partial}{\partial z^B_{(1)}} - \frac{\partial \xi^B_{(3)}}{\partial z^A_{(2)}} \frac{\partial}{\partial z^B_{(2)}} - \frac{\partial \xi^B_{(4)}}{\partial z^A_{(2)}} \frac{\partial}{\partial z^B_{(3)}},$$

$$J_1[\xi, \frac{\partial}{\partial z^A_{(2)}}] = - \frac{\partial \xi^B_{(2)}}{\partial z^B_{(2)}} \frac{\partial}{\partial z^B_{(2)}} - \frac{\partial \xi^B_{(3)}}{\partial z^A_{(2)}} \frac{\partial}{\partial z^B_{(3)}}$$

$$d_T d_{J_2} L(\frac{\partial}{\partial z^A_{(2)}}) = 0,$$

$$d_T d_{J_2} L(\xi) = z^B_{(1)} \, d_T(\frac{\partial L}{\partial z^B_{(2)}}) + \xi^B_{(2)} \frac{\partial L}{\partial z^B_{(2)}}$$

and

$$d_T d_{J_2} L[\xi, \frac{\partial}{\partial z^A_{(2)}}] = - \frac{\partial \xi^B_{(2)}}{\partial z^A_{(2)}} \frac{\partial L}{\partial z^B_{(2)}}.$$

Therefore, (15) becomes

$$z^B_{(1)} \frac{\partial^2 L}{\partial z^B_{(1)} \partial z^A_{(2)}} + \frac{1}{2} \xi^B_{(2)} \frac{\partial^2 L}{\partial z^B_{(2)} \partial z^A_{(2)}} - \frac{1}{2} z^B_{(1)} \frac{\partial}{\partial z^A_{(2)}} (d_T(\frac{\partial L}{\partial z^B_{(2)}}))$$

and, on the other hand, (16) becomes

$$z^B_{(1)} \frac{\partial^2 L}{\partial z^B_{(1)} \partial z^A_{(2)}} + z^B_{(2)} \frac{\partial^2 L}{\partial z^B_{(2)} \partial z^A_{(2)}} - \frac{1}{2} z^B_{(1)} \frac{\partial}{\partial z^A_{(2)}} (d_T(\frac{\partial L}{\partial z^B_{(2)}})).$$

Consequently, from (12) and the regularity of L, we obtain $\xi^B_{(2)} = 2z^B_{(2)}$, $1 \le B \le m$, that is, ξ is a semispray of type 2.

Step 3: Take $X = \frac{\partial}{\partial z^A_{(1)}}$. Since $\xi = z^B_{(1)} \frac{\partial}{\partial z^B_{(0)}} + 2z^B_{(2)} \frac{\partial}{\partial z^B_{(1)}}$ $+ \xi^B_{(3)} \frac{\partial}{\partial z^B_{(2)}} + \xi^B_{(4)} \frac{\partial}{\partial z^B_{(3)}}$, we have

$$[\xi, \frac{\partial}{\partial z^A_{(1)}}] = -\frac{\partial}{\partial z^B_{(0)}} - \frac{\partial \xi^B_{(3)}}{\partial z^A_{(1)}} \frac{\partial}{\partial z^B_{(2)}} - \frac{\partial \xi^B_{(4)}}{\partial z^A_{(1)}} \frac{\partial}{\partial z^B_{(3)}},$$

$$J_1[\xi, \frac{\partial}{\partial z^A_{(1)}}] = -\frac{\partial}{\partial z^B_{(1)}} - \frac{\partial \xi^B_{(3)}}{\partial z^A_{(1)}} \frac{\partial}{\partial z^B_{(3)}},$$

$$d_T d_{J_2} L(\frac{\partial}{\partial z^A_{(1)}}) = \frac{\partial L}{\partial z^A_{(2)}}$$

$$d_T d_{J_2} L(\xi) = z^B_{(1)}\, d_T(\frac{\partial L}{\partial z^B_{(2)}}) + 2z^B_{(2)} \frac{\partial L}{\partial z^B_{(2)}},$$

and

$$d_T d_{J_2} L[\xi, \frac{\partial}{\partial z^A_{(1)}}] = -d_T(\frac{\partial L}{\partial z^A_{(2)}}).$$

Now, from (12) and proceeding as above, we have $\xi^B_{(3)} \frac{\partial^2 L}{\partial z^B_{(2)} \partial z^A_{(2)}} = 3\, z^B_{(3)} \frac{\partial^2 L}{\partial z^B_{(2)} \partial z^A_{(2)}}$, and, taking into account that L is regular, we deduce $\xi^B_{(3)} = 3\, z^B_{(3)}$, $1 \leq B \leq m$. Hence ξ is a semispray of type 1. This concludes the proof of Proposition (6). □

The next theorem shows that (12) is, in fact, the intrinsic form of the Lagrangian equations of motion.

THEOREM (2). Let $L: T^k M \to \mathbb{R}$ be a regular Lagrangian of order k on M and ξ the semispray on $T^{2k-1}M$ of type 1 given by (12). Then the paths of ξ satisfy the Euler-Lagrange equations

$$\frac{\partial L}{\partial q^A} - \frac{d}{dt}\frac{\partial L}{\partial \dot{q}^A} + \frac{d^2}{dt^2}\frac{\partial L}{\partial \ddot{q}^A} + \dots + (-1)^k \frac{d^k}{dt^k}\frac{\partial L}{\partial q^{(k)A}} = 0,$$

$A = 1,\dots,m.$

Proof: We take $k = 2$ (the proof for an arbitrary k is obtained from a similar procedure and it is left to the reader as an exercise). In this case, (10) and (11) reduce to $\omega_L = -dd_{J_1}L + \frac{1}{2}\, dd_T d_{J_2} L$ and $E = C_1 L - \frac{1}{2}\, d_T(C_2 L) - L$, respectively. Let ξ be the vector field given by (12). From Proposition (6), ξ is a semispray on T^3M of type 1; then, ξ is locally expressed by $\xi = z^B_{(1)}\frac{\partial}{\partial z^B_{(0)}} + 2z^B_{(2)}\frac{\partial}{\partial z^B_{(1)}} + 3z^B_{(3)}\frac{\partial}{\partial z^B_{(2)}} + \xi^B\frac{\partial}{\partial z^B_{(3)}}$, where $\xi^B = \xi^B(z^C_{(0)}, z^C_{(1)}, z^C_{(2)}, z^C_{(3)})$.

Hence, we have

$$[\xi, \frac{\partial}{\partial z^A_{(0)}}] = -\frac{\partial \xi^B}{\partial z^A_{(0)}}\frac{\partial}{\partial z^B_{(3)}},$$

$$J_1[\xi, \frac{\partial}{\partial z^A_{(0)}}] = 0,$$

$$d_T d_{J_2} L(\frac{\partial}{\partial z^A_{(0)}}) = d_T(\frac{\partial L}{\partial z^A_{(2)}}),$$

$$d_T d_{J_2} L(\xi) = z^B_{(1)} d_T(\frac{\partial L}{\partial z^B_{(2)}}) + 2z^B_{(2)}\frac{\partial L}{\partial z^B_{(2)}} = d_T(C_2 L),$$

and

$$d_T d_{J_2} L[\xi, \frac{\partial}{\partial z^A_{(0)}}] = 0,$$

since $C_2 L = z^B_{(1)}\frac{\partial L}{\partial z^B_{(2)}}$.

Then, from (15), we have

$$(i_\xi \omega_L)\left(\frac{\partial}{\partial z^A_{(0)}}\right) = -\xi\left(\frac{\partial L}{\partial z^A_{(1)}}\right) + \frac{\partial}{\partial z^A_{(0)}}(C_1 L)$$

$$+ \frac{1}{2}\xi\left(d_T\left(\frac{\partial L}{\partial z^A_{(2)}}\right)\right) - \frac{1}{2}\frac{\partial}{\partial z^A_{(0)}}(d_T(C_2 L)), \qquad (15)'$$

and, from (16), we have

$$dE\left(\frac{\partial}{\partial z^A_{(0)}}\right) = \frac{\partial}{\partial z^A_{(0)}}(C_1 L) - \frac{1}{2}\frac{\partial}{\partial z^A_{(0)}}(d_T(C_2 L)) - \frac{\partial L}{\partial z^A_{(0)}}. \qquad (16)'$$

From the comparison of (15)' and (16)', onde deduces that

$$\frac{\partial L}{\partial z^A_{(0)}} - \xi\left(\frac{\partial L}{\partial z^A_{(1)}}\right) + \frac{1}{2}\xi\left(d_T\left(\frac{\partial L}{\partial z^A_{(2)}}\right)\right) = 0. \qquad (17)$$

Now, let σ be a path of ξ. Then, along the prolongation $\tilde{\sigma}^3$ of σ to T^3M, the equation (17) becomes

$$\frac{\partial L}{\partial q^A} - \frac{d}{dt}\frac{\partial L}{\partial \dot{q}^A} + \frac{d^2}{dt^2}\frac{\partial L}{\partial \ddot{q}^A} = 0,$$

since $\frac{1}{2} d_T\left(\frac{\partial L}{\partial z^A_{(2)}}\right) = \frac{d}{dt}\frac{\partial L}{\partial \ddot{q}^A}$, along $\tilde{\sigma}^3$. □

(b). - <u>The Lagrangian</u> L <u>is irregular</u>

In this case, the degeneracy of L implies that ω_L will now be merely presymplectic. Consequently, the equation (12) will not in general posses globally defined solutions and even if solutions exist they will not be unique.

Then, we suppose that there is on $T^{2k-1}M$ an almost product structure P such that the following conditions are satisfied:

(i) (ω_L, P) is a horizontal presymplectic structure on $T^{2k-1}M$, and,

(ii) $dE \in \operatorname{Im} P^*$,

where E is the energy of L defined by (11).

From Proposition (5), one easily deduces

PROPOSITION (7). <u>With the hypothesis above, there is only one vector field</u> $\xi \in \operatorname{Im} P$ <u>satisfying</u> (12), <u>that is</u>,

$$i_\xi \omega_L = dE.$$

REMARK (10). An alternative procedure was proposed by Gotay and Nester (1979)(a), (1979)(b), (see Appendix B). They have developed in the case $k = 1$, a gometric constraint algorithm which gives necessary and sufficient conditions for the solvability of "consistent" Lagrangian equations of motion. Specifically, the algorithm finds whether or not there exists a submanifold S of TM along which (12) holds; if such a manifold exists, then the algorithm gives a constructive method for finding it. The algorithm of Gotay and Nester can be applied to every presymplectic manifold; then, if we consider the presymplectic manifold $(T^{2k-1}M, \omega_L)$ the algorithm yields a submanifold S of $T^{2k-1}M$ along which (12) holds, if such a manifold exists. Actually, both approaches are related as follows: If S is any integral manifold of the distribution defined by $\operatorname{Im} P$ on $T^{2k-1}M$, then (12) holds along S and the integral curves of ξ lies on S.

As Gotay and Nester (1979)(a) remark in the case $k=1$, ξ will not necessarily be a semispray (or a second order equation). Consequently, the condition $J_1\xi = C_1$ must be imposed in order to assume that the paths of ξ satisfy the Euler-Lagrange equations corresponding to L. Then, we have

THEOREM (3). Let L, ω_L, P, ξ be as above and suppose that ξ is a semispray of type 1 on $T^{2k-1}M$. Then the paths of ξ are the solutions of the Euler-Lagrange equations

$$\frac{\partial L}{\partial q^A} - \frac{d}{dt}\frac{\partial L}{\partial \dot{q}^A} + \frac{d^2}{dt^2}\frac{\partial L}{\partial \ddot{q}^A} + \ldots + (-1)^k \frac{d^k}{dt^k}\frac{\partial L}{\partial q^{(k)A}} = 0,$$

$A = 1,\ldots,m$, $\dim M = m$.

Proof: The Theorem (3) can be proved in the same way as Theorem (2) and its verification is left to the reader. □

REMARK (11). Gotay and Nester prove that, if $k = 1$ and L possesses an additional structure of "admissibility", then exists a submanifold $\tilde{S}$ of S and a vector field ξ on $\tilde{S}$ such that ξ satisfies (12) and it is a semispray, where S is the submanifold of TM given by the algorithm.

2.4. Homogeneous Lagrangians

In this section, we develop the generalized Klein's formalism for homogeneous Lagrangians. The interest of this study lies on the following facts: If the Lagrangian L is homogeneous, then the vector field given by (12) is a spray; consequently, we can associate to ξ a homogeneous connection Γ whose geodesics are precisely the solutions of the Euler-Lagrange equations corresponding to L.

We distinguish two cases. If the Lagrangian L is regular, then the result follows directly, but, if the Lagrangian L is irregular, then additional conditions on the horizontal presymplectic structure are needed in order to assure that ξ is a spray.

We propose the following definition.

DEFINITION (3). Let L be a Lagrangian of order k on M, that is, $L: T^kM \to \mathbb{R}$ is a function on the tangent bundle of order k of M. Then L is said to be <u>homogeneous</u> if the functions $L, C_2L, C_3L, \dots, C_kL$ are homogeneous of degree $2, 1, 0, \dots, 3-k$, respectively.

From Definition (3), one easily deduces

PROPOSITION (8). <u>If the Lagrangian</u> L <u>is homogeneous, then the energy</u> E <u>of</u> L, <u>given by</u> (11), <u>is homogeneous of degree</u> 2 <u>and the 2-form</u> ω_L, <u>given by</u> (10), <u>is homogeneous of degree</u> 1.

<u>Proof</u>: In fact, if L is homogeneous, then L, $d_T(C_2L)$, $d_T^2(C_3L)$, ..., $d_T^{k-1}(C_kL)$ are homogeneous of degree 2, because of Proposition (4). Consequently, we have

$$\begin{aligned} C_1E &= C_1L - \frac{1}{2!} C_1(d_T(C_2L)) + \frac{1}{3!} C_1(d_T^2(C_3L)) \\ &\quad + \dots + (-1)^{k-1} \frac{1}{k!} C_1(d_T^{k-1}(C_kL)) \\ &= 2L - \frac{1}{2!} 2d_T(C_2L) + \frac{1}{3!} 2d_T^2(C_3L) \\ &\quad + \dots + (-1)^{k-1} \frac{1}{k!} 2(d_T^{k-1}(C_kL)) \\ &= 2(L - \frac{1}{2!} d_T(C_2L) + \frac{1}{3!} d_T^2(C_3L) + \dots + (-1)^{k-1} \frac{1}{k!} d_T^{k-1}(C_kL)) \\ &= 2E, \end{aligned}$$

and, so, E is homogeneous of degree 2.

In a similar way, we can prove that ω_L is homogeneous of degree 1. □

Now, we can state the main result for regular Lagrangians.

PROPOSITION (9). *Let* L *be a homogeneous regular Lagrangian of order* k *on* M. *Then the unique vector field* ξ *on* $T^{2k-1}M$ *giving by (12) is a spray of type 1.*

Proof: As it is well known, we have

$$i_{[C_1,\xi]}\omega_L = \mathcal{L}_{C_1} i_\xi \omega_L - i_\xi \mathcal{L}_{C_1}\omega_L .$$

But $i_\xi \omega_L = dE$, and, consequently,

$$\begin{aligned}\mathcal{L}_{C_1} i_\xi \omega_L &= \mathcal{L}_{C_1} dE \\ &= d\,\mathcal{L}_{C_1} E \\ &= 2dE \\ &= 2\, i_\xi \omega_L ,\end{aligned}$$

since E is homogeneous of degree 2 by Proposition (8). Then, one has that

$$i_{[C_1,\xi]}\omega_L = i_\xi \omega_L ,$$

because ω_L is homogeneous of degree 1, again, by Proposition (8).

Hence, the non-degeneracy of ω_L implies that $[C_1,\xi] = \xi$. Consequently, ξ is a homogeneous vector field on $T^{2k-1}M$ of degree 1. Since ξ is a semispray of type 1 by Proposition (6), the result follows. □

Now, we consider an irregular Lagrangian L of order k on M. As we remark in the beginning of this section, the homogeneity of L does not imply that the vector field ξ given by (12) is homogeneous. Nevertheless, we have the following.

PROPOSITION (9). *Let* L *be a homogeneous irregular Lagran-*

gian of order k on M such that the horizontal presymplectic structure (ω_L, P) satisfies the following conditions:

(i) *The distribution Im P is involutive (that is, for all vector fields* $X, Y \in \mathrm{Im}\, P$, $[X,Y] \in \mathrm{Im}\, P$), *and*

(ii) $C_1 \in \mathrm{Im}\, P$.

Then the unique vector field $\xi \in \mathrm{Im}\, P$ *given by* (12) *is homogeneous of degree* 2.

Proof: As in the regular case, the homogeneity of L implies that

$$i_{[C_1,\xi]}\omega_L = i_\xi \omega_L .$$

But Im P is involutive, then $[C_1,\xi] \in \mathrm{Im}\, P$ and, consequently,

$$[C_1,\xi] = \xi ,$$

taking into account Proposition (5). □

COROLLARY (1). *If, in addition,* ξ *is a semispray of type* 1, *then* ξ *is a spray (of type* 1).

REMARK (12). We can express the Corollary (1) geometrically as follows. From Frobenius theorem, Im P defines a foliation on $T^{2k-1}M$ such that ω_L restricted on each leaf of this foliation is non degenerate. Consequently, each leaf becomes a symplectic manifold and the integral curves of the spray ξ (that is, the dynamical trajectories corresponding to L) lie on it.

Let now $\langle\ ,\ \rangle$ be a Riemannian metric on a manifold M and let $L: TM \to \mathbb{R}$ given by

$$L(v) = \frac{1}{2}\langle v,v\rangle .$$

In local coordinates $(q^A, \dot{q}^A)$, L is given by

$$L(q^A, \dot{q}^A) = \frac{1}{2} g_{AB} \dot{q}^A \dot{q}^B,$$

where $g_{AB} = \langle \frac{\partial}{\partial \dot{q}^A}, \frac{\partial}{\partial \dot{q}^B} \rangle$ is the matrix of the metric $\langle\ ,\ \rangle$. Obviously, L is homogeneous of degree 2 with respect to the velocities $\dot{q}^A$ and the velocity Hessian of L is

$$\left(\frac{\partial^2 L}{\partial \dot{q}^A \partial \dot{q}^B}\right) = (g_{AB}).$$

Hence L is a homogeneous regular Lagrangian of order 1 on M and it is easy to check that the spray ξ given by (12) is locally given by

$$\xi = \dot{q}^A \frac{\partial}{\partial q^A} - \Gamma^A_{BC} \dot{q}^B \dot{q}^C \frac{\partial}{\partial \dot{q}^A}, \tag{18}$$

where $\Gamma^A_{BC} = \frac{1}{2} g^{AD} \left(\frac{\partial g_{DC}}{\partial q^B} - \frac{\partial g_{DB}}{\partial q^C} - \frac{\partial g_{BC}}{\partial q^D}\right)$ are the Christoffel symbols and (g^{AD}) is the inverse matrix of (g_{AD}).

From (18), we see that the geodesics of ξ are the geodesics of the Lévi-Civita connection ∇ defined by $\langle\ ,\ \rangle$ (Abraham & Marsden (1978)). Moreover, ξ is the spray canonically associated to ∇ according to §3.9 of Chap. I and which will called the <u>geodesic spray</u>.

To end this section, we extend this situation to any homogeneous Lagrangian of higher order, regular or not. Indeed, if L is a homogeneous Lagrangian of order k on a manifold M, then we can associate to L a homogeneous connection Γ of type 1 on $T^{2k-1}M$ in such a way that the geodesic of Γ are the solutions of the Euler-Lagrange equations corresponding to L.

In fact, let $L: T^kM \to \mathbb{R}$ be a homogeneous Lagrangian of order k on M. If we suppose that L is regular or otherwise, that L satisfies the hypothesis of Corollary (1), then the vector field ξ given by (12) is a spray on $T^{2k-1}M$ of type 1.

Then, from Proposition (34), §3.9, of Chap. I, we have

THEOREM (4). The connection Γ of type 1 on $T^{2k-1}M$ given by

$$\Gamma = \frac{1}{2k}\{-2[\xi, J_1] + (2k-2)\mathrm{Id}\}$$

is homogeneous and its associated spray is precisely ξ.

COROLLARY (2). Γ and ξ have the same geodesics. Consequently, the geodesics of Γ are the solutions of the Euler-Lagrange equations of L.

Proof: The proof follows from Proposition (34), §3.9, Chap. I, and Theorems (2) and (3).

II.3. - Generalized Hamiltonian formalism

3.1. Jacobi-Ostrogradsky coordinates and Legendre transformation

First, we shall show that the differential calculus on higher tangent bundles developed above allow us to define, in a canonical way, the Jacobi-Ostrogradsky momentum coordinates corresponding to a Lagrangian of higher order.

Indeed, let L be a Lagrangian of order k on a manifold M. Then we can define a Pfaff form α_L on $T^{2k-1}M$ by

$$\alpha_L = d_{J_1}L - \frac{1}{2!} d_T d_{J_2} L + \frac{1}{3!} d_T^2 d_{J_3} L$$
$$+\ldots+ (-1)^{k-1} \frac{1}{k!} d_T^{k-1} d_{J_k} L \,. \qquad (1)$$

From (1), it is easy to prove that α_L is locally given by

$$\alpha_L = \sum_{s=1}^{k} (s-1)!\, f_{A/s}\, dz^A_{(s-1)} \,, \qquad (2)$$

$$f_{A/s} = \sum_{i=0}^{k-s} (-1)^i \frac{1}{(s+i)!} d_T^i \left(\frac{\partial L}{\partial z^A_{(s+i)}}\right) . \qquad (3)$$

For instance, if $k = 2$ we have a Pfaff form $\alpha_L = d_J L$ on TM locally given by

$$\alpha_L = \frac{\partial L}{\partial \dot{q}^A} dq^A \qquad \text{(see Godbillon (1969))}$$

and, if $k = 2$, a Pfaff form $\alpha_L = d_{J_1}L - \frac{1}{2} d_T d_{J_2} L$ on T^3M locally given by

$$\alpha_L = \left(\frac{\partial L}{\partial z^A_{(1)}} - \frac{1}{2} d_T\left(\frac{\partial L}{\partial z^A_{(2)}}\right)\right) dz^A_{(0)} + \frac{1}{2} \frac{\partial L}{\partial z^A_{(2)}} dz^A_{(1)} \,.$$

Let us return to the general case. Since a Pfaff form β on $T^{2k-1}M$ is semibasic of type r if and only if β is locally given by

$$\beta = \sum_{i=0}^{2k-r-1} \beta_A(z^A_{(0)},\ldots,z^A_{(2k-1)})\, dz^A_{(i)} \,,$$

we deduce, from (2), the following.

PROPOSITION (1). α_L <u>is a semibasic Pfaff from on</u> $T^{2k-1}M$ <u>of type</u> k.

Now, let σ be a curve in M and $\tilde{\sigma}^{2k-1}$ its prolongation to $T^{2k-1}M$. A simple calculation from (3) shows that, along $\tilde{\sigma}^{2k-1}$, $f_{A/s}$ becomes

$$p_{A/s} = \sum_{i=0}^{k-s} (-1)^i \frac{d^i}{dt^i} \frac{\partial L}{\partial q^{(s+i)A}} , \tag{4}$$

where $1 \le s \le k$. Then α_L, along $\tilde{\sigma}^{2k-1}$, becomes

$$\alpha_L = \sum_{s=1}^{k} p_{A/s}\, dq^{(s-1)A} . \tag{5}$$

We adopt the following.

DEFINITION (1). The Pfaff form α_L will be called the <u>Jacobi-Ostrogradsky form</u> corresponding to the Lagrangian L.

Now, taking into account Theorem (1), §3.6, Chap. I and Proposition (1), there exists a mapping $\text{Leg}: T^{2k-1}M \to T^*(T^{k-1}M)$ such that

$$q_{T^{k-1}M} \circ \text{Leg} = \rho^{2k-1}_{k-1} ,$$

that is, the following diagram

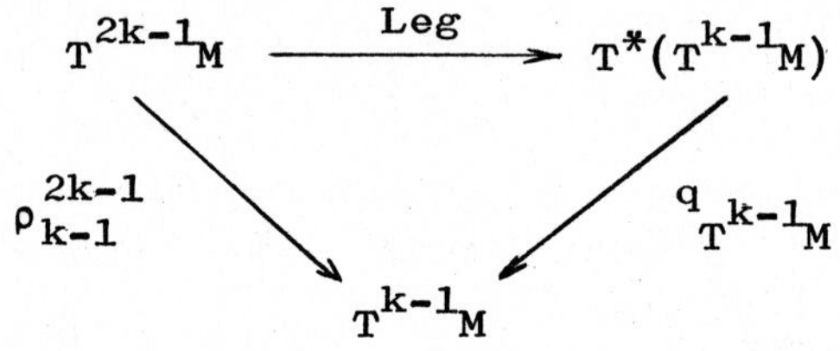

is commutative. Moreover, if λ_{k-1} is the Liouville form on $T^*(T^{k-1}M)$, then we have

$$\text{Leg}^* \lambda_{k-1} = \alpha_L , \tag{6}$$

because Proposition (24), §3.6, Chap. I. Then, if $\omega_{k-1} = -d\lambda_{k-1}$ is the canonical symplectic form on $T^*(T^{k-1}M)$, we have

$$\text{Leg}^* \omega_{k-1} = \omega_L , \tag{7}$$

because $\omega_L = -d\alpha_L$.

From (26), §3.6, Chap. I and (2), we deduce that Leg is locally given by

$$(z^A_{(0)},\dots,z^A_{(2k-1)}) \rightsquigarrow (z^A_{(0)},\dots,z^A_{(k-1)},f_{A/1},f_{A/2},2!f_{A/3},\dots,(k-1)!f_{A/k}), \quad (8)$$

and, consequently, if σ is a curve in M, (8) becomes

$$(q^A,\dot{q}^A,\dots,q^{(2k-1)s}) \rightsquigarrow (q^A,\dots,q^{(k-1)A},p_{A/1},\dots,p_{A/k}), \quad (8)'$$

along the prolongation $\tilde{\sigma}^{2k-1}$ of σ to $T^{2k-1}M$, taking into account (4).

REMARK (1). Actually, (7) justifies the choice of ω_L in (10), §3.3.

In the case $k = 1$, we know that the Lagrangian L is regular if and only if the mapping Leg given by (8) is a local diffeomorphism (see Godbillon (1969) and Abraham & Marsden (1978)) and, in such a case, Leg is the Legendre transformation corresponding to L. For an arbitrary k this result is still valid. In fact, we have

THEOREM (1). The following assertions are equivalent:

(i) L is regular

(ii) ω_L is a symplectic form

(iii) Leg: $T^{2k-1}M \to T^*(T^{k-1}M)$ is a local diffeomorphism.

In such a case, the mapping Leg will be called the "Legendre transformation" corresponding to L.

Proof: The equivalence between (i) and (ii) has been established in §3.3. Then we shall prove the equivalence between (ii) and (iii). Indeed, from (7) and (8) we have

$$\omega_L = \sum_{s=1}^{k} (s-1)!\, dz^A_{(s-1)} \wedge df_{A/s} \,. \quad (9)$$

Now, the result follows directly from (9). □

From Theorem (1), we deduce that the Lagrangian L is regular if and only if the sets of functions $(z^A_{(0)},\dots,z^A_{(k-1)},f_{A/1},\dots,(k-1)!f_{A/k})$ are coordinate systems. In such a case, (9) shows that $(z^A_{(0)},\dots,z^A_{(k-1)},f_{A/1},\dots,$ $(k-1)!f_{A/k})$ are canonical coordinates with respect to ω_L. Then we adopt the following definition.

DEFINITION (2). Let L be a regular Lagrangian of order k on a manifold M. Then the set of coordinates in $T^{2k-1}M$

$$(z^A_{(0)},\dots,z^A_{(k-1)},f_{A/1},\dots,(k-1)!f_{A/k})$$

will be called the <u>Jacobi-Ostrogradsky momentum coordinates</u>. When a $(2k-1)$ jet prolongation is involved we will consider the set of coordinates

$$(q^A,\dot{q}^A,\dots,q^{(k-1)A},p_{A/1},\dots,p_{A/k}).$$

The Legendre transformation Leg corresponding to a regular Lagrangian L is not necessarily a diffeomorphism. If Leg is in fact a difeomorphism, then L is said to be <u>hyperregular</u>. Since $q_{T^{k-1}M} \circ \mathrm{Leg} = \rho^{2k-1}_{k-1}$, Leg is a diffeomorphism if and only if its restriction to each fibre of $T^{2k-1}M \xrightarrow{\rho^{2k-1}_{k-1}} T^{k-1}M$ is one-to-one.

Now, let L be a hyperregular Lagrangian of order k on M with energy E given by (11), §3.3, and ξ the vector field on $T^{2k-1}M$ given by (12), §3.3, that is, $i_\xi\omega_L = dE$. Then the vector field $\xi^* = T\,\mathrm{Leg} \circ \xi \circ \mathrm{Leg}^{-1}$ on $T^*(T^{k-1}M)$ is given by

$$i_{\xi^*}\omega_{k-1} = d(E\circ Leg^{-1}).$$

Indeed,

$$\begin{aligned} i_{\xi^*}\omega_{k-1} &= i_{\xi^*}(Leg^{-1})^*\omega_L \\ &= (Leg^{-1})^* i_\xi \omega_L \\ &= (Leg^{-1})^* dE \\ &= d(E\circ Leg^{-1}). \end{aligned}$$

Hence, we have

PROPOSITION (2). *If* σ *is an integral curve of* ξ, *then* $\gamma = Leg\circ\sigma$ *is an integral curve of* ξ^* *and we have* $q_{T^{k-1}M}\circ\gamma = \rho^{2k-1}_{k-1}\circ\sigma$.

In the sequel, the energy function E of a regular Lagrangian L (and the function $E\circ Leg^{-1}$ on $T^*(T^{k-1}M)$ if L is in addition hyperregular) will be called the *Hamiltonian energy* corresponding to L and denoted by H.

3.2. Hamilton equations

The introduction of the Jacobi-Ostrogradsky momentum coordinates permits to develop the *Hamiltonian form* of the generalized Euler-Lagrange dynamical equations. In order to do this, we shall work in local coordinates. Obviously, the canonical equations can be also derived if one considers the Hamiltonian system $(T^{2k-1}M, \omega_L, H)$ if L is regular (or $(T^*(T^{k-1}M),\omega_{k-1},H)$ if L is hyperregular) and applies the usual formalism described in the introduction of this chapter.

The generalized Euler-Lagrange equations along jet prolongations of curves in M are

$$\frac{\partial L}{\partial q^A} - \frac{d}{dt}\frac{\partial L}{\partial \dot{q}^A} + \frac{d^2}{dt^2}\frac{\partial L}{\partial \ddot{q}^A} + \dots + (-1)^k \frac{d^k}{dt^k}\frac{\partial L}{\partial q^{(k)A}} = 0,$$

and the Hamiltonian energy corresponding to such (regular) Lagrangian L is

$$H = \sum_{s=1}^{k} p_{A/s}\, q^{(s)A} - L(q^A, \dot{q}^A, \dots, q^{(k)A}),$$

where

$$p_{A/s} = \sum_{i=0}^{k-s} (-1)^i \frac{d^i}{dt^i}\frac{\partial L}{\partial q^{(s+i)A}}, \qquad 1 \le s \le k,$$

are the Jacobi-Ostrogradsky momentum coordinates. The comparison of the differentials

$$dH = \frac{\partial H}{\partial q^A} dq^A + \sum_{s=1}^{k} \frac{\partial H}{\partial p_{A/s}} dp_{A/s} + \sum_{s=1}^{k-1} \frac{\partial H}{\partial q^{(s)A}} dq^{(s)A}$$

and

$$dH = -\frac{\partial L}{\partial q^A} dq^A + \sum_{s=1}^{k} q^{(s)A} dp_{A/s} + \sum_{s=1}^{k} \left(p_{A/s} - \frac{\partial L}{\partial q^{(s)A}}\right) dq^{(s)A}$$

shows that

$$\frac{\partial H}{\partial q^A} = -\frac{\partial L}{\partial q^A} = -\frac{d}{dt} p_{A/1} \qquad \text{(see the Euler-Lagrange equations and the definition of } p_{A/1})$$

$$\frac{\partial H}{\partial p_{A/s}} = q^{(s)A} = \frac{d}{dt} q^{(s-1)A}, \qquad \text{with} \quad 1 \le s \le k.$$

On the other hand, we have

$$\begin{aligned} p_{A/s} - \frac{\partial L}{\partial q^{(s)A}} &= \sum_{i=0}^{k-s} (-1)^i \frac{d^i}{dt^i}\frac{\partial L}{\partial q^{(s+i)A}} - \frac{\partial L}{\partial q^{(s)A}} \\ &= \sum_{i=1}^{k-s} (-1)^i \frac{d^i}{dt^i}\frac{\partial L}{\partial q^{(s+i)A}} \\ &= \sum_{i=0}^{k-(s+1)} (-1)^{i+1} \frac{d}{dt}\left(\frac{d^i}{dt^i}\frac{\partial L}{\partial q^{(s+1+i)A}}\right) \\ &= -\frac{d}{dt} p_{A/s+1}, \qquad 1 \le s \le k-1. \end{aligned}$$

Therefore

$$\frac{\partial H}{\partial q^{(s)A}} = -\frac{d}{dt} p_{A/s+1}, \qquad 1 \le s \le k-1$$

and we have the equations

$$\frac{\partial H}{\partial q^{(s-1)A}} = -\frac{d}{dt} p_{A/s} ; \quad \frac{\partial H}{\partial p_{A/s}} = \frac{d}{dt} q^{(s-1)A}, \quad 1 \le s \le k, \qquad (10)$$

which are the <u>canonical form</u> of the generalized Euler-Lagrange dynamical equations.

REMARK (2). The canonical equations (10) can be directly obtained from (3), §1, taking into account (9) in Theorem (1). Actually, the reader can easily check that the Legendre transformation Leg corresponding to L and given by (8)' permits to pass from the Euler-Lagrange equations to the equations (10) (this fact is a direct consequence of Proposition (2) if the Lagrangian L is hyperregular).

REMARK (3). We may express the equations (10) in a more symmetrical form:

$$\frac{\partial H}{\partial q^{(s)A}} = -\frac{d}{dt} p_{A/s} ; \quad \frac{\partial H}{\partial p_{A/s}} = \frac{d}{dt} q^{(s)A}, \text{ with } 0 \le s \le k-1. \qquad (10)'$$

To do this, we adopt the following convention:

$$p_{A/-1} = \frac{\partial L}{\partial q^A} - \frac{d}{dt}\frac{\partial L}{\partial \dot{q}^A} + \ldots + (-1)^k \frac{d^k}{dt^k}\frac{\partial L}{\partial q^{(k)A}} = 0$$

(the Euler-Lagrange equation)

and

$$p_{A/s} = \sum_{i=0}^{k-(s+1)} (-1)^i \frac{d^i}{dt^i}\frac{\partial L}{\partial q^{(s+1+i)A}}.$$

In such a case the Hamiltonian is defined by

$$H = \sum_{s=0}^{k-s} p_{A/s}\, q^{(s+1)A} - L.$$

3.3. Variational approach

It is well known that Lagrangian theories arise from variational principles. The discussion of such interesting problem in the geometrical scope will be presented in Chapter III. There we will comment the relation between higher order variational problems in many independent variables and the zero order variational problems. More clearly, it is known that in the usual situation where only one independent variable is involved, there exists an equivalence between variational problems generated by functionals depending on regular Lagrangians of order s (called zero order variational problem) and "modified" variational problems given by functionals depending on the Poincaré-Cartan invariant form (called order one variational problem). This equivalence is characterized by an injective/surjective relation between the sets of extremals of the respective problems. The main question is to know whether this result holds for variational problems of order $k > 1$ with several independent variables.

In a note published in 1979, P. Dedecker studied such a problem for higher order theories in one independent variable. Let L be a Lagrangian of order k on $\mathbb{R} \times T^kM$ and consider the 1-form

$$\Omega = L\, dt + \sum_{s=1}^{k} p_{A/s}(dq^{(s-1)A} - q^{(s)A}dt), \tag{11}$$

along the jet prolongation of a curve σ in $\mathbb{R}\times M$, where the p's are the Jacobi-Ostrogradsky momenta. With the help of the definition of the Hamiltonian we may re-write Ω as:

$$\Omega = \sum_{s=1}^{k} p_{A/s}\, dq^{(s-1)A} - H\, dt, \qquad (11)'$$

which is called the <u>generalized Poincaré-Cartan form</u>.

REMARK (4). Actually, the reader can easily see that the generalized Poincaré-Cartan form Ω is nothing but the 1-form $\alpha_L - H\,dt$ on $\mathbb{R} \times T^kM$.

The main result of Dedecker's note is the following.

THEOREM (2) (The generalized Helmholtz-Cartan theorem).
If the Hessian matrix

$$\frac{\partial^2 L}{\partial q^{(k)A} \partial q^{(k)B}}$$

is invertible, then the variational problem of order k

$$\delta\, I_k(\sigma) = \delta \int_\sigma L(t, q^A, \dot{q}^A, \ldots, q^{(k)A})dt = 0,$$

is equivalent to the variational problem of order zero

$$\delta\, I_o = \delta \int_{\tilde{\sigma}^{2k-1}} \Omega = 0,$$

where σ *is a curve in* $\mathbb{R} \times M$. *This equivalence is characterized by the fact that the extremals of* I_o *are solutions of the equations*

$$dq^{(i-1)A} - q^{(i)A}\, dt = 0, \qquad 1 \le i \le 2k-1,$$

and also of the generalized Euler-Lagrange equations and reciprocally.

(For a clearer interpretation of this last equality see Proposition (2), §3.2, Chap. III.)

The regular condition on L assures the <u>local</u> invertibility of the transformation

$$(t,q^A,\dots,q^{(k-1)A},\dots,q^{(2k-1)A}) \rightsquigarrow (t,q^A,\dots,q^{(k-1)A},p_{A/1},\dots,p_{A/k}), \quad (12)$$

where the p's are the Jacobi-Ostrogradsky momenta.

For the autonomous situation, it is now clear that a <u>Lagrangian $L: T^kM \to \mathbb{R}$ is regular if and only if it verifies the Helmholtz-Cartan regular condition.</u> Consequently, (12) is nothing but the change of coordinates given by the Legendre transformation corresponding to L.

The local expression of the Poincaré-Cartan form along jet prolongations is given by (11) (or (11)'). Therefore, from (9), we have

$$d\Omega = \sum_{s=1}^{k} dp_{A/s} \wedge dq^{(s-1)A} - dH \wedge dt$$

$$= -\omega_L - dH \wedge dt.$$

For the autonomous situation and regular Lagrangians,

$$d\Omega = -\omega_L$$

is a symplectic form on $T^{2k-1}M$ (recall that $\dim T^{2k-1}M = 2k\cdot\dim M$). The symplectic form of the generalized Hamilton equations is

$$i_{X_H}\omega_L = dH$$

and therefore the integrals of X_H verify the Hamilton equations (10). For the non-autonomous situation the Helmholtz-Cartan generalized theorem assures that "the set of extremals is a foliation of dimension one on $T^{2k-1}M$, the 1-form Ω induces a contact structure and the differential $d\Omega$ defines a symplectic structure on the leaf space" (Dedecker (1979)).

Therefore, we have

THEOREM (3) (Dedecker (1979)). The form Ω (resp., $d\Omega$) is a relative (resp., absolute) invariant integral of the extremals of $\int \Omega$.

3.4. Poisson brackets, canonical transformations and Hamilton-Jacobi equation

In what follows we suppose that a regular Lagrangian $L: T^k M \to \mathbb{R}$ is given, a curve σ in M is prolongated to $T^{2k-1}M$ and the Jacobi-Ostrogradsky coordinates are taken.

DEFINITION (3). The Poisson brackets of two functions f and g on $T^{2k-1}M$ are defined by

$$\{f,g\}_{q^{(s-1)A},\,p_{A/s}} = \sum_{A=1}^{m} \sum_{s=1}^{k} \left(\frac{\partial f}{\partial q^{(s-1)A}} \frac{\partial g}{\partial p_{A/s}} - \frac{\partial f}{\partial p_{A/s}} \frac{\partial g}{\partial q^{(s-1)A}}\right), \quad (13)$$

where

$$\{f,g\}_{q^{(s-1)A},\,p_{A/s}} = \{f,g\}_{(q^A,\ldots,q^{(k-1)A},\,p_{A/1},\ldots,p_{A/k})}.$$

REMARK (5). The reader can easily check that

$$\{f,g\}_{q^{(s-1)A},\,p_{A/s}} = \omega_L(X_f,X_g),$$

where X_f and X_g are the vector fiels on $T^{2k-1}M$ given by $i_{X_f}\omega_L = df$ and $i_{X_g}\omega_L = dg$, respectively.

A few properties, similar to those found for Poisson brackets in Classical Mechanics, can be easily proved:

(i) $\{q^{(j-1)A},q^{(i-1)B}\} = 0$; (ii) $\{p_{A/j},p_{B/i}\} = 0$

(iii) $\{q^{(j-1)A},p_{B/i}\} = \delta^A_B\,\delta^{j-1}_i$.

One can also express equations of motion (10) with the

aid of Poisson brackets

$$\{q^{(s-1)B},H\}_{q^{(i-1)A},p_{A/i}} =$$

$$= \sum_{A=1}^{m}\sum_{i=1}^{k}\left(\frac{\partial q^{(s-1)B}}{\partial q^{(i-1)A}}\frac{\partial H}{\partial p_{A/i}} - \frac{\partial q^{(s-1)B}}{\partial p_{A/i}}\frac{\partial H}{\partial q^{(i-1)A}}\right)$$

$$= \sum_{A=1}^{m}\sum_{i=1}^{k}\left(\frac{\partial q^{(s-1)B}}{\partial q^{(i-1)A}}\frac{\partial H}{\partial p_{A/i}}\right)$$

$$= \frac{\partial H}{\partial p_{B/s}}$$

$$= \frac{d}{dt}\, q^{(s-1)B} .$$

Similarly,

$$\{H,p_{B/s}\} = -\frac{d}{dt}\, p_{B/s} .$$

In standard Classical Mechanics, a <u>canonical transformation</u> is a mapping of the phasespace onto itself which preserves the form of the canonical equations of motion. This property is a consequence of the preservation of the symplectic form under the action of such mapping. In our presente generalized Mechanics we have a closed 2-form ω_L on $T^{2k-1}M$ which may be symplectic if we suppose that L is regular. Therefore, we will say that a mapping $\varphi : T^{2k-1}M \to T^{2k-1}M$ is canonical if ω_L is invariant under the action of such a mapping.

We now study a necessary and sufficient condition for a mapping to be canonical. In order to do this, we examine a change of coordinates for the autonomous case:

$$(q^{(i-1)A},p_{A/i}) \rightsquigarrow (Q^{(j-1)B},p_{B/j}), \qquad i,j = 1,\ldots,k$$
$$A,B = 1,\ldots,m.$$

PROPOSITION (3). A necessary and sufficient condition for the above change of coordinates to be canonical is that

$$\frac{\partial q^{(i-1)A}}{\partial Q^{(j-1)B}} = \frac{\partial P_{B/j}}{\partial p_{A/i}} ; \qquad \frac{\partial q^{(i-1)A}}{\partial P_{B/j}} = - \frac{\partial Q^{(j-1)B}}{\partial p_{A/i}}$$

$$\frac{\partial p_{A/i}}{\partial Q^{(j-1)B}} = - \frac{\partial P_{B/j}}{\partial q^{(i-1)A}} ; \qquad \frac{\partial p_{A/i}}{\partial P_{B/j}} = \frac{\partial Q^{(j-1)B}}{\partial q^{(i-1)A}}$$

Proof: We want a (time-independent) transformation to lead us from the set of variables q's and p's to another set where the variables are the Q's and P's, and such that

$$H(q^{(i-1)A},p_{A/i}) = K(Q^{(j-1)B}(q^{(i-1)A},p_{A/i}),P_{B/j}(q^{(i-1)A},p_{A/i}))$$

$$\left.\begin{aligned} \dot{Q}^{(j-1)B} &= \frac{d}{dt} Q^{(j-1)B} = \frac{\partial K}{\partial P_{B/j}} \\ \dot{P}_{B/j} &= \frac{d}{dt} P_{B/j} = - \frac{\partial K}{\partial Q^{(j-1)B}} \end{aligned}\right\} \quad \text{(Hamilton equations)}$$

The symplectic form of the Hamilton equations is, for each Hamiltonian,

$$i_{X_H} \omega_L = dH ; \qquad i_{X_K} \bar{\omega}_L = dK,$$

where $\bar{\omega}_L$ is ω_L in the new coordinates.

Let us suppose that the change of coordinates is canonical. Then $X_H = X_K$ and locally one has

$$X_H = \frac{\partial H}{\partial p_{A/i}} \frac{\partial}{\partial q^{(i-1)A}} - \frac{\partial H}{\partial q^{(i-1)A}} \frac{\partial}{\partial p_{A/i}} = \dot{q}^{(i-1)A} \frac{\partial}{\partial q^{(i-1)A}} + \\ + \dot{p}_{A/i} \frac{\partial}{\partial p_{A/i}}$$

$$X_H(q^{(i-1)A}) = \dot{q}^{(i-1)A} = \dot{Q}^{(j-1)B} \frac{\partial q^{(i-1)A}}{\partial Q^{(j-1)A}} + \dot{P}_{B/j} \frac{\partial q^{(i-1)A}}{\partial P_{B/j}} =$$

$$= X_K(q^{(i-1)A}) \tag{*}$$

$$X_H(p_{A/i}) = \dot{p}_{A/i} = \dot{Q}^{(j-1)B} \frac{\partial p_{A/i}}{\partial Q^{(j-1)B}} + \dot{P}_{B/j} \frac{\partial p_{A/i}}{\partial P_{B/j}} =$$

$$= X_K(p_{A/i}). \tag{**}$$

On the other hand,

$$X_H = \left(\frac{\partial K}{\partial Q^{(j-1)B}} \frac{\partial Q^{(j-1)B}}{\partial p_{A/i}} + \frac{\partial K}{\partial P_{B/j}} \frac{\partial P_{B/j}}{\partial p_{A/i}}\right) \frac{\partial}{\partial q^{(i-1)A}} -$$

$$- \left(\frac{\partial K}{\partial Q^{(j-1)B}} \frac{\partial Q^{(j-1)B}}{\partial q^{(i-1)A}} + \frac{\partial K}{\partial P_{B/j}} \frac{\partial P_{B/j}}{\partial q^{(i-1)A}}\right) \frac{\partial}{\partial p_{A/i}}$$

$$= \left(- \frac{\partial Q^{(j-1)B}}{\partial p_{A/i}} \dot{P}_{B/j} + \frac{\partial P_{B/j}}{\partial p_{A/i}} \dot{Q}^{(j-1)B}\right) \frac{\partial}{\partial q^{(i-1)A}} -$$

$$- \left(- \frac{\partial Q^{(j-1)B}}{\partial q^{(i-1)A}} \dot{P}_{B/j} + \frac{\partial P_{B/j}}{\partial q^{(i-1)A}} \dot{Q}^{(j-1)B}\right) \frac{\partial}{\partial p_{A/i}} .$$

The substitution of (*) and (**) into X_H and the comparison with the above expression gives the desired result. We leave to the reader the proof in the other direction. □

COROLLARY. <u>The above change of coordinates is canonical if and only if</u>

$$\{F,G\}_{q^{(i-1)A}, p_{A/i}} = \{F,G\}_{Q^{(j-1)B}, P_{B/j}} .$$

<u>Proof</u>: A direct calculation shows that

$$\{F,G\}_{q^{(i-1)A}, p_{A/i}} =$$

$$= \sum_{A,B=1}^{m} \sum_{i,j=1}^{k} \left\{ \frac{\partial F}{\partial Q^{(j-1)B}} \left(\frac{\partial G}{\partial p_{A/i}} \frac{\partial p_{A/i}}{\partial P_{B/j}} + \frac{\partial G}{\partial q^{(i-1)A}} \frac{\partial q^{(i-1)A}}{\partial P_{B/j}}\right) - \right.$$

$$- \frac{\partial F}{\partial P_{B/j}} \left(\frac{\partial G}{\partial p_{A/i}} \frac{\partial p_{A/i}}{\partial Q^{(j-1)B}} + \frac{\partial G}{\partial q^{(i-1)A}} \frac{\partial q^{(i-1)A}}{\partial Q^{(j-1)B}} \right) \}$$

$$= \sum_{B=1}^{m} \sum_{j=1}^{k} \left(\frac{\partial F}{\partial Q^{(j-1)B}} \frac{\partial Q^{(j-1)B}}{\partial P_{B/j}} - \frac{\partial F}{\partial P_{B/J}} \frac{\partial G}{\partial Q^{(j-1)B}} \right) = \{F,G\}_{Q^{(j-1)B},P_{B/j}} .$$

□

To find the Hamilton-Jacobi equation for this Mechanics, we must define the generator function of the canonical transformations.

Let us consider a system describable by two sets of variables such that

$$\delta \int (p_{A/i}\, q^{(i)A} - H)dt = 0$$

$$\delta \int (P_{A/i}\, Q^{(i)A} - K)dt = 0$$

The following expression must hold:

$$(p_{A/i} dq^{(i-1)A} - Hdt) - (P_{A/i} dQ^{(i-1)A} - Kdt) = dG. \qquad (14)$$

If we choose $G = G(q^{(i-1)A}, Q^{(i-1)A}, t)$ and if it is explicitly developed then the comparison of the corresponding coefficients gives

$$p_{A/i} = \frac{\partial G}{\partial q^{(i-1)A}}, \quad P_{A/i} = - \frac{\partial G}{\partial Q^{(i-1)A}}, \quad K = H + \frac{\partial G}{\partial t} \qquad (15)$$

and G is the generator function of a canonical transformation.

A generator function S, whose $Q^{(i-1)A}$ and $P_{A/i}$ are constants, gives:

$$\partial K/\partial P_{A/i} = (d/dt)Q^{(i-1)A} = 0$$

$$\partial K/\partial Q^{(i-1)A} = -(d/dt)P_{A/i} = 0$$

$$dK/dt = \partial K/\partial t \ .$$

From these equations it is possible to impose $\partial K/\partial t = 0$: K will then be a constant which may be regarded as zero. We obtain, accordingly, from (14), $(\partial S/\partial t) + H(q^{(i-1)A}, p_{A/i}, t) = 0$, or, using (15), $(\partial S/\partial t) + H(q^{(i-1)A}, (\partial S/\partial q^{(i-1)A}, t)) = 0$, which is our <u>Hamilton-Jacobi equation</u>.

II.4. - Exercises

1. Prove (ii) of Proposition (3), §2.1.

2. Prove the following identity:

$$[d_{J_r}, d_T] = d_{J_r} d_T - d_T d_{J_r} = \begin{cases} d & , \text{ if } r = 1 \\ r d_{J_{r-1}} & , \text{ if } r \geq 2. \end{cases}$$

3. Prove (6), §2.1, for an arbitrary k.

4. (i) Prove that

$$i_{J_o} = (i_{J_1} - \frac{1}{2!} d_T i_{J_2} + \frac{1}{3!} d_T^2 i_{J_3} + \ldots) d_T$$

(Hint: Use Proposition (2)).

(ii) Prove that $\delta d = 0$ and $\delta d_T = 0$. Then prove that $\delta^2 = 0$.

(iii) Prove that if u is a form such that $\delta u = 0$, then there exists two local forms v and w such that $u = dv + d_T w$.

(iv) A 1-form u such that $\delta u = 0$ and $i_{J_1} u = 0$ will be called a <u>Euler-Poisson form</u>. Prove that if L is a function in Λ, then δL is a Euler-Poisson form.

(v) (Poincaré lemma). Let u be a Euler-Poisson form. Prove that there exists a (local) function L in Λ such that $u = \delta L$ (Hint: Use (iii)). (For more details, see Tulczyjew (1975)(a)).

5. Prove Theorem (2), §2.3, for an arbitrary k.

6. Let L be a Lagrangian of order 2 on $M = \mathbb{R}$ given by $L(x,y,z) = yz + z^2$.

   (i) Compute ω_L and check that it is a symplectic form on $T^3\mathbb{R} = \mathbb{R}^4$.

   (ii) Compute the energy E of L and the vector field ξ given by $i_\xi \omega_L = dE$. Then check that ξ is a semi-spray on T^3M of type 1.

   (iii) Check that the paths of ξ are in fact the solutions of the Euler-Lagrange equations corresponding to L.

   (iv) Write the Jacobi-Ostrogradsky form, the Legendre transformation and the canonical equations corresponding to L.

CHAPTER III

GENERALIZED FIELD THEORY

III.1 - Introduction

Generalized Particle Mechanics was developed for a time-independent (or autonomous) formalism on the tangent bundle of order k with one (implicit) variable t, the functions $y^A(t)$ and their time derivatives up to order k. We want now to consider a more general situation involving a certain number of independent variables x_a, a set of (smooth) functions $y^A(x_a)$ and their partial derivatives up to a given order k with respect to all x_a. We are interested in an extension of the theory which preserve, as far as possible, the standard field formalism.

We will see in the forthcoming paragraphs that the major difficulty concerns the geometric formulation of the variational theory in terms of the intrinsical Hamilton-Cartan procedure. Even at the local level there is no unique way of putting the Euler-Lagrange equations for higher order Lagrangians in their canonical form. The formalism presents some problems. Among then we can mention the possibility of having different definitions of the Legendre transformation due to the concept of non-degeneracy of the Lagrangian and the ambiguities in the understanding the physical meaning of "canonical conjugate variables" (see for example Thielheim (1967) and Coelho de Souza and Rodrigues (1969)). The definition of the "generalized momentum variables" is an intriguing problem not only for mathematicians but also physicists and different approaches are found in the literature. For example, Chang (1946), (1948) proposed a theory where the canonical equations are derived without an explicit definition of the momenta; in de Wett's (1948) alternative approach, the Hamiltonian formalism does not contain the usual one as a particular case.

Our study starts with the local "traditional" formulation of Variational Calculus in a real Euclidean space. We consider two types of variations for the action principle: one compares the transformed variables (which in our theory include terms involving higher derivatives) and the original ones at the same point; the other includes also the variation of the independent variables (§2.2 and §2.3). From this it is possible to present the Hamiltonization in two forms: the first, given in §2.4, may be considered as an extension of Jacobi-Ostrogradsky formalism in a parametric form and the second, presented in §2.5, may be considered as a covariant formulation of Jacobi-Ostrogradsky-de Donder theory (Cf. de Donder (1935)). This last paragraph is inspired in the article of Musicki (1978). The classical topics on canonical transformations and Poisson brackets is presented in §2.6. We leave to the reader the development of Hamilton-Jacobi theory for both situations (see also Musicki (1978)). Some comments about Noether's theorem are presented at the end of the chapter.

Next we try to put the formalism in terms of the Jet bundle theory. We consider the Lagrangians as real smooth functions defined on the manifold J^kM of k-jets of (eventually local) sections of a vector fibered manifold (M,p,N), with M being trivial. The integer k is finite and arbitrarily chosen and the manifold N is locally parametrized by the (x_a) and M by (x_a, y^A). Therefore the Lagrangians $\mathfrak{L}: J^kM \to \mathbb{R}$ are functions of type $\mathfrak{L} = \mathfrak{L}(x_a, y^A, z^A_{a_1}, \ldots, z^A_{a_1 \ldots a_k})$, $1 \leq a_1 \leq \ldots \leq a_k$, with the z's being also coordinates for J^kM.

Unfortunately the geometrical formalism for variational theory in terms of exterior differential forms for higher order theories is complicated and its <u>global</u> formulation does not preserve the same results of the first order theory. More clearly, the intrinsical formalism does not maintain the unicity of classical Poincaré-Cartan integral invariant of Analytical Mechanics. In this standard situation we know that Hamilton's principle may be presented either by a first order variational problem in one independent variable given by the integral

$$I_1 = \int L(t, q_a, dq_a/dt)dt, \qquad (*)$$

or by a zero-th order variational problem in one independent variable given by the integral of the one-form

$$\Omega = L(t, q_a, dq_a/dt)dt + (\partial L/\partial q_a)\theta_a, \quad \theta_a = dq_a - v_a dt.$$

This one-form is uniquely determined and is called Poincaré-Cartan. For regular Lagrangians there is an injective-surjective mapping between the set of the extremals of both variational problems. When we try to extend this result for the k-th order variational problems in n-independent variables, that is, when we extend (*) to Lagrangians $\mathfrak{L}$ of order k, we will deal with integrals of type

$$I_k = \int (\tilde{s}^k)^* \mathfrak{L}\omega,$$

where ω is a volume form on N injected into J^kM and $\tilde{s}^k$ is the k-jet prolongation of a section s of (M,p,N). The major problem now is to show that there is a n-form Ω for which it is possible to define a zero-th order variational problem equivalent (in the above sense) to the original integral. This problem has interested many people in this last years. It is known that if $k > 1$ and $n = 1$ or if $k = 1$ and $n > 1$ then there is a unique (Poincaré-Cartan) form for wich both variational problems are equivalent for the regular case. When $k > 1$ and $n > 1$ then it is possible to show that the Poincaré-Cartan form is not uniquely determined (see the approaches of Garcia & Muñoz (1983) and Horák & Kolár (1983)). This suggests a new interpretation on the above mentioned problem on the Hamiltonization of higher order Lagrangians.

The discussion of such interesting problem in its global formulation will not be examined in detail in the text, but we will make comments and indications for the reader.

In §3.4 we present two local attempts in the Hamiltonian direction: one gives a formulation on the dual manifold of $J^1(J^{k-1}M)$ with the help of constraints and the other is developed over the jet manifold $J^{2k-1}M$. We propose in §3.5

a way to relate them with aid of the so-called Dirac-Bergmann theory of constraints, injecting $J^{2k-1}M$ into $J^1(J^{2k-2}M)$. Some final considerations and examples are given at the end of the chapter and we think that such few examples suggest future investigations.

The formalism adopted in this chapter is the usual one (and we recall that Generalized Particle Mechanics can be obtained as a particular case when we take $N = \mathbb{R}$). We recall that the "almost tangent" procedure may be also adopted for J^kM (See Remark (2), sect. I.3, Ch.I). Finally, we would like to stress that the problem of "qualifying" correctly the conjugate momenta for higher order Hamiltonians in the physical sense remains a question to be decided by the nature of the problem studied. In one way or the other the "aesthetic" mathematical value of higher order field theories justifies the work to obtain them.

III.2. - Traditional method formalism

2.1 Higher order Lagrangian densities.

It is known that whenever the Lagrangian formulation of a field theory is concerned, the local dynamics is obtained from a smooth function $\mathfrak{L}$, called Lagrangian density. In the absence of gravitacional forces, this density is a function depending on objects of the 4-dimensional Minkowsky space-time manifold, to each point of which are assigned the coordinates x_a, where a takes the values from 1 to 4: the field variables $y^A = y^A(x_a)$, $1 \leq A \leq m$, which are supposed smooth functions of the space-time coordinates and their derivatives $y_a^A = (\partial y^A/\partial x_a)$. So $\mathfrak{L} = \mathfrak{L}(x_a, y^A, y_a^A)$. The functional Lagrangian L is defined as

$$L = \int_D \mathfrak{L}\, dx_1 dx_2 dx_3 dx_4 = \int_D \mathfrak{L}\, d^4x,$$

where D is some (compact) domain in the Minkowsky space-time manifold. The variational principle for such functional leds to the Euler-Lagrange field equations

$$\frac{\partial \mathfrak{L}}{\partial y^A} - \Sigma_a \frac{d}{dx_a} \left(\frac{\partial \mathfrak{L}}{\partial y_a^A}\right) = 0. \tag{1}$$

Equations (1) may be written in the following form:

$$\frac{\partial \mathfrak{L}}{\partial y^A} - \frac{d}{dx_1} \left(\frac{\partial \mathfrak{L}}{\partial y_1^A}\right) - \sum_2^3{}_b \frac{d}{dx_b} \left(\frac{\partial \mathfrak{L}}{\partial y_b^A}\right) = 0.$$

Introducing the functional (or variational) derivative of L with respect to y^A

$$\frac{\delta L}{\delta y^A} = \frac{\partial \mathfrak{L}}{\partial y^A} - \Sigma_b \frac{d}{dx_b} \left(\frac{\partial \mathfrak{L}}{\partial y_b^A}\right),$$

and to y_b^A

$$\frac{\delta L}{\delta y_b^A} = \frac{\partial \mathfrak{L}}{\partial y_b^A} - \Sigma_c \frac{d}{dx_c} \left(\frac{\partial \mathfrak{L}}{\partial y_{bc}^A}\right);$$

taking into account that second order derivatives of the fields do not appear in the Lagrangian, one has

$$\frac{\delta L}{\delta y^A} - \frac{d}{dx_1} \left(\frac{\delta L}{\delta y_1^A}\right) = 0 \tag*{(1)$_1$}$$

which gives a "parametric" version of equations (1) for an arbitrary choice of coordinates (x_a). Equations $(1)_1$ show some resemblance with the Euler-Lagrange equations in Particle Mechanics.

A generalization of the above description involving higher order Lagrangians starts with a smooth function (the higher order Lagrangian density) $\mathfrak{L}: \mathbb{R}^K \to \mathbb{R}$, where K is given by

$$K = n+m+n\cdot m+(1/2)(n(n+1)m) + \ldots + \binom{k+n-1}{n-1}m; \qquad (*)$$

$\binom{a}{b} = \frac{a!}{(a-b)!b!}$ and the density $\mathfrak{L}$ depends on $\underline{n}$ independent variables (x_a), $\underline{m}$ smooth function $y^A(x_a)$ and their partial derivatives with respect to x_a up to order k. So

$$\mathfrak{L} = \mathfrak{L}(x_a, y^A, \frac{\partial y^A}{\partial x_a}, \ldots, \frac{\partial^k y^A}{\partial x_{a_1} \ldots \partial x_{a_k}}). \qquad (2)$$

REMARK (1). A more general situation may be considered if the Lagrangian density is a function of higher derivatives of different orders with respect to the independent variables. For example, let i be a partition of the set of integers $\{1,2,\ldots,n\}$ and j the complementary. We may, for instance, have a Lagrangian density of type

$$\mathfrak{L} = \mathfrak{L}(x_i, x_j, y^A(x_i, x_j), \frac{\partial y^A}{\partial x_i}, \ldots, \frac{\partial^s y^A}{\partial x_{i_1} \ldots \partial x_{i_s}},$$
$$\frac{\partial^\ell y^A}{\partial x_{j_1} \ldots \partial x_{j_\ell}}, \ldots, \frac{\partial^{k+\ell} y^A}{\partial x_{i_1} \ldots \partial x_{i_k} \partial x_{j_1} \ldots \partial x_{j_\ell}})$$

with $1 \le s \le k$, $k \le \ell \le r$. We will not consider such a situation here, but we shall return to this point in III.4.

Let us return to the Lagrangian density (2). We want to obtain the corresponding generalized Euler-Lagrange field equations for this theory. As we know, this involves many different problems in Variational Calculus since it depends on what kind of extremals are considered for a given functional. As Ω contains higher order derivatives the class of variations for which the action is stationary might be further restricted and it is not evident what the further restrictions are. As we are not interested in discussing this kind of problems we will consider only two types of variations, generalizing to higher dimensions the method described, for example, in the classical book of Gelfand and Fomin (1963), p.152 and sequents.

For sake of simplicity, throughout the text we will assume the following conventions:

$$x = (x_a) = (x_1,\dots,x_n); \quad d^n x = dx_1 \dots dx_n; \quad d^{n-1}x = dx_2 \dots dx_n, \quad \text{etc}\dots$$

$$y = (y^A) = (y^1,\dots,y^m)$$

$$y^A_{a(s)} = y^A_{a_1\dots a_s} = \frac{\partial^s y^A}{\partial x_{a_1}\dots\partial x_{a_s}} \quad \text{for} \quad 1 \le a_1 \le \dots \le a_s \le n \quad \text{and} \quad 1 \le s \le k.$$

2.2. <u>Higher order Euler-Lagrange field equations</u>.

Let us first consider the situation where the variation of the functional is taken in such a way that the region D of integration is compact (we restrict ourselves, for example, to the case where D is a "hyperparallelepiped" P^n) and such that the (infenitesimal) transformation, for a small ε, given by

$$\left.\begin{aligned} y^A \mapsto Y^A &= y^A + \varepsilon f^A + \text{higher powers of } \varepsilon \\ &\vdots \\ y^A_{a(s)} \mapsto Y^A_{a(s)} &= y^A_{a(s)} + \varepsilon f^A_{a(s)} + \text{higher powers of } \varepsilon \end{aligned}\right\} \quad (3)$$

remains in D, while the coordinates x_a remain fixed for all a.

DEFINITION (1). Let $\mathcal{L}: \mathbb{R}^k \to \mathbb{R}$ be a smooth function. By the variation δJ of the functional

$$J(y) = \int_{P^n} \mathcal{L}(x_a, y^A, y^A_{a(s)}) d^n x \tag{4}$$

with respect to the transformation (3), we mean the term of order one in ε of $J(Y) - J(y)$;

$$\delta J = \int_{P^n} [\mathcal{L}(x, y^A + \varepsilon f^A, y^A_{a(s)} + \varepsilon f^A_{a(s)}) - \mathcal{L}(x, y^A, y^A_{a(s)})] d^n x.$$

THEOREM (1). The higher order Euler-Lagrange field equations are

$$\frac{\partial \mathcal{L}}{\partial y^A} - \Sigma_{a_1} \frac{d}{dx_{a_1}} \left(\frac{\partial \mathcal{L}}{\partial y^A_{a_1}}\right) + \ldots +$$

$$(-1)^k \Sigma_{a_1 \ldots a_k} \frac{d^k}{dx_{a_1} \ldots dx_{a_k}} \left(\frac{\partial L}{\partial y^A_{a_1 \ldots a_k}}\right) = 0 \tag{5}$$

provided that $f^A = f^A_{a_1} = \ldots = f^A_{a_1 \ldots a_k} = 0$ in the boundary of the respective domain of integration.

Proof: The proof is a generalization of the standard procedure for $k = 1$. Using Taylor's theorem, we find the variation of (4) to be

$$\delta J = \varepsilon \int_{P^n} \left\{\frac{\partial \mathcal{L}}{\partial y^A} f^A + \Sigma_s \Sigma_{a(s)} \frac{\partial \mathcal{L}}{\partial y^A_{a(s)}}\right\} d^n x. \tag{6}$$

Integrating by parts and using Green's theorem, we get:

$$\delta J = \epsilon \int_{P^n} \{[\frac{\partial \mathcal{L}}{\partial y^A} + \Sigma_s (-1)^s \Sigma_{a(s)} \frac{d^s}{dx_{a_1} \ldots dx_{a_s}} (\frac{\partial \mathcal{L}}{\partial y^A_{a(s)}})] f^A\} d^n x$$

$$+ \epsilon \{ [\int_{P^{n-1}} \Sigma_{a(1)} (\frac{\partial \mathcal{L}}{\partial y^A_{a(1)}} f^A) d^{n-1} x] +$$

$$+ [\int_{P^{n-2}} \Sigma_{a(2)} (\frac{\partial \mathcal{L}}{\partial y^A_{a(2)}} f^A_{a(1)} - \frac{d}{dx_{a_2}} \frac{\partial \mathcal{L}}{\partial y^A_{a(2)}} f^A) d^{n-2} x] + \ldots$$

$$+ [\Sigma_{a(k)} (\frac{\partial \mathcal{L}}{\partial y^A_{a(k)}} f^A_{a(k-1)} + \ldots + (-1)^{k-1} \frac{d^{k-1}}{dx_{a_2} \ldots dx_{a_k}} \frac{\partial \mathcal{L}}{\partial y^A_{a(k)}} f^A]\},$$

where we have used

$$\frac{\partial \mathcal{L}}{\partial y^A_{a(1)}} f^A_{a(1)} = \frac{d}{dx_{a_1}} (\frac{\partial \mathcal{L}}{\partial y^A_{a(1)}} f^A) - \frac{d}{dx_{a_1}} \frac{\partial \mathcal{L}}{\partial y^A_{a(1)}} f^A$$

$$\frac{\partial \mathcal{L}}{\partial y^A_{a(2)}} f^A_{a(2)} = \frac{d}{dx_{a_2}} (\frac{\partial \mathcal{L}}{\partial y^A_{a(2)}} f^A_{a(1)}) - \frac{d}{dx_{a_2}} \frac{\partial \mathcal{L}}{\partial y^A_{a(2)}} f^A_{a(1)}$$

$$= \frac{d}{dx_{a_2}} (\frac{\partial \mathcal{L}}{\partial y^A_{a(2)}} f^A_{a(1)}) - \frac{d}{dx_{a_1}} (\frac{d}{dx_{a_2}} \frac{\partial \mathcal{L}}{\partial y^A_{a(2)}} f^A) + \frac{d^2}{dx_{a_1} dx_{a_2}} \frac{\partial \mathcal{L}}{\partial y^A_{a_1 a_2}} f^A,$$

etc, and where $d^{n-1}x, d^{n-2}x, \ldots,$ are the volume elements in the respective subdomains of $\mathbb{R}^n$. It is now clear that the conditions $f^A = \ldots = f^A_{a_1 \ldots a_{k-1}} = 0$ lead to equations (5).

□

We have seen above that the Euler-Lagrange equations may be expressed in terms of functional derivatives. For the present situation if we put

$$\frac{\delta \Omega}{\delta y^A_{a(1)}} = \frac{\partial \Omega}{\partial y^A_{a(1)}} - \Sigma_{a_2} \frac{d}{dx_{a_2}} \left(\frac{\partial \Omega}{\partial y^A_{a(2)}}\right) + \dots + (-1)^k \Sigma_{a_2 \dots a_k} \frac{d^{k-1}}{dx_{a_2} \dots dx_{a_k}} \left(\frac{\partial \Omega}{\partial y^A_{a(k)}}\right),$$

$$\vdots$$

$$\frac{\delta \Omega}{\delta y^A_{a(s)}} = \frac{\partial \Omega}{\partial y^A_{a(s)}} - \Sigma_{a_{s+1}} \frac{d}{dx_{a_{s+1}}} \left(\frac{\partial \Omega}{\partial y^A_{a(s+1)}}\right) + \dots + (-1)^k \Sigma_{a_{s+1} \dots a_k} \frac{d^{k-s}}{dx_{a_{s+1}} \dots dx_{a_k}} \left(\frac{\partial \Omega}{\partial y^A_{a(k)}}\right)$$

$$\vdots$$

$$\frac{\delta \Omega}{\delta y^A_{a(k)}} = \frac{\partial \Omega}{\partial y^A_{a(k)}}, \tag{7}$$

then (5) takes the concise form

$$\frac{\delta \Omega}{\delta y^A} = 0. \tag{5$_1$}$$

2.3. Functionals defined on a variable domain.

We have derived the Euler-Lagrange equations supposing that all x's variables remain fixed. Let us now suppose that the variation is total. We start with a functional J defined on a domain D and a transformation $T: (x_a, y^A, y^A_{a(s)}) \rightsquigarrow (\bar{x}_a, \bar{y}^A, \bar{y}^A_{a(s)})$ defined by

$$\bar{x}_a = x_a + \varepsilon g^a; \qquad \bar{y}^A = y^A + \varepsilon f^A$$

$$\bar{y}^A_{a(s)} = y^A_{a(s)} + \varepsilon f^A_{a(s)}$$

(only the linear part of ε is considered). A more especific study may be performed. Barut and Mullen (1962)(b), for example, considered infinitesimal translation plus rotations, (this article should be referred not only for this question, but also for a similar treatment for higher order field theories). We summarize the results in the following way: consider the functional

$$J = \int_D \Omega(x_a, y^A, y^A_{a(s)})\, d^n x$$

and the variation δJ produced by the action of the above transformation T. Let us put

$$\left.\begin{aligned} \delta(x_a) &= \bar{x}_a - x_a = \varepsilon g^a \\ \delta(y^A) &= \bar{y}^A - y^A = \varepsilon f^A ; \\ \delta(y^A_{a(s)}) &= \bar{y}^A_{a(s)}(\bar{x}_a) - y^A_{a(s)}(x_a) = \varepsilon f^A_{a(s)} . \end{aligned}\right\} \qquad (8)$$

We must also consider the variations

$$\bar{\delta} y^A = \bar{y}^A(x_a) - y^A(x_a); \qquad \bar{\delta} y^A_{a(s)} = \bar{y}^A_{a(s)}(x_a) - y^A_{a(s)}(x_a),$$

corresponding to the action of T along the x-coordinates. We introduce new functions $\bar{f}$ by

$$\bar{\delta} y^A = \varepsilon \bar{f}^A, \qquad \bar{\delta} y^A_{a(s)} = \varepsilon \bar{f}^A_{a(s)}$$

related to the old f by

$$\bar{f}^A = f^A - y^A_{a_1} g^{a_1}, \ldots, \bar{f}^A_{a(s)} = f^A_{a(s)a_{s+1}} g^{a_{s+1}} .$$

It can be shown that the total variation of J corresponding to the transformation T is given by the formula (Gelfand & Fomin (1963), p.173)

$$\delta J = \varepsilon \int_D \frac{\delta J}{\delta y^A} \bar{f}^A d^n x +$$

$$+ \varepsilon \int_D \Big(\sum_0^{k-1} r \frac{d}{dx_{a_{r+1}}} \Big(\frac{\delta J}{\delta y^A_{a(r+1)}} \bar{f}^A_{a(r)} + \Omega g^{a_{r+1}} \Big) d^n x, \qquad (9)$$

where, for conciseness, we have put $f^A_{a(0)} = f^A$, $y^A_{a(0)} = y^A$.

For fixed boundary conditions we obtain the above field

equations $\delta J/\delta y^A = 0$ and so for extremals, (9) reduces to

$$\delta J = \int_D \sum_0^{k-1} r \frac{d}{dx_{a_{r+1}}} \left(\frac{\delta J}{\delta y^A_{a(r+1)}} (\varepsilon \bar{f}^A_{a(r)}) + \mathcal{L}(\varepsilon g^{a_{r+1}})\right) d^n x ,$$

or, from the definition of T and (6),

$$\delta J = \int_D \sum_0^{k-1} r \frac{d}{dx_{a_{r+1}}} \left[\frac{\delta J}{\delta y^A_{a(r+1)}} \delta y^A_{a(r)} - \left(\frac{\delta J}{\delta y^A_{a(r+1)}} y^A_{a(r)a_{r+1}} - \mathcal{L}\right) \delta x_{a_{r+1}}\right] d^n x. \quad (9)_1$$

2.4. An extension of Jacobi-Ostrogradsky's approach.

THEOREM (2) (Jacobi-Ostrogradsky's approach for Field Theory). There exists an appropriate equivalent canonical formulation for (regular) Lagrangian theories involving derivatives order higher than one.

Proof: In order to obtain such formulation let us single out one variable, say x_1. It will be useful to introduce the following generalization of functional derivative:

DEFINITION (2). Let $f: \mathbb{R}^k \to \mathbb{R}$ be a smooth function. The r-functional (or variational) derivative $(\delta_r F/\delta_r y^A)$ of the functional

$$F = \int f(x_a, y^A, y^A_{a(s)}) d^{n-1}x$$

with respect to y^A is

$$\frac{\delta_r F}{\delta_r y^A} = \frac{\partial f}{\partial y^A} - \sum_2^n a(1) \frac{d}{dx_{a_1}} \left(\frac{\partial f}{\partial y^A_{a(1)}}\right) + \ldots +$$

$$(-1)^r \sum_2^n a(r) \frac{d^r}{dx_{a_1} \ldots dx_{a_r}} \left(\frac{\partial f}{\partial y^A_{a(r)}}\right) , \quad (10)$$

where $1 \leq r \leq k$ (recall that all summations start from 2).

It is clear that we can also "derive" F with respect to the other variables and that

$$\frac{\delta_o F}{\delta_o y^A} = \frac{\partial f}{\partial y^A}, \dots, \frac{\delta_o F}{\delta_o y^A_{a(k)}} = \frac{\partial f}{\partial y^A_{a(k)}}.$$

Also, for all r,

$$\frac{\delta_o F}{\delta_r x_a} = \frac{\partial f}{\partial x_a}.$$

With the help of this definition equations (5) may be written as

$$\frac{\delta_k L}{\delta_k y^A} - \frac{d}{dx_1}\left(\frac{\delta_{k-1} L}{\delta_{k-1} y^A_1}\right) + \dots + (-1)^k \frac{d^k}{dx_1^k}\left(\frac{\delta_o L}{\delta_o y^A_{\underbrace{1\dots1}_{k \text{ times}}}}\right) = 0 \qquad (5)_2$$

where now

$$L = \int \mathcal{L}\, dx_2 \dots dx_n = \int \mathcal{L}\, d^{n-1}x,$$

(if we put $J = \int L\, dx_1$, then we have $(5)_2$).

Let us introduce the following notation:

$$y^A_{(\ell)} = \frac{d^\ell y^A}{dx_1^\ell}, \qquad 0 \le \ell \le k, \qquad (y^A_o \overset{\text{def}}{=} y^A).$$

Then $(5)_2$ takes the form

$$\sum_{\ell=0}^{k} (-1)^\ell \frac{d^\ell}{dx_1^\ell}\left(\frac{\delta_{k-\ell} L}{\delta_{k-\ell} y^A_{(\ell)}}\right) = 0, \qquad (5)_3$$

which shows a very close formal resemblance with the equations

$$\sum_{\ell=0}^{k} (-1)^\ell \frac{d^\ell}{dt^\ell}\left(\frac{\partial L}{\partial q^i_{(\ell)}}\right) = 0$$

describying, as we have seen before, Generalized Mechanical Systems including higher time derivatives of the coordinates

$q^i(t)$. By analogy, we define the <u>conjugate momentum of higher order</u> as

$$p_{A/s} = \sum_{0}^{k-s}{}_{\ell} (-1)^{\ell} \frac{d^{\ell}}{dx_1^{\ell}} \left(\frac{\delta_{k-\ell-s} L}{\delta_{k-\ell-s} y^A_{(\ell+s)}}\right) \tag{11}$$

with $1 \leq s \leq k$. It is clear that $p_{A/0} = 0$, since for $s = 0$ we have $(5)_3$.

We are now able to define the <u>Hamiltonian</u> for the present approach as a functional

$$H(x_1, y^A, \ldots, y^A_{(k-1)}, p_{A/1}, \ldots, p_{A/k}) = \int h \, d^{n-1}x,$$

where the <u>Hamiltonian density</u> h is defined by

$$h = \Sigma_A \, \Sigma_{\ell} \, p_{A/\ell} \, y^A_{(\ell)} - \mathfrak{L} \tag{12}$$

and we finish the proof of the theorem with the following

PROPOSITION (1). <u>The Hamiltonian form of the Euler-Lagrange equations</u> (5) <u>is</u>

$$\frac{\delta_{k-s} H}{\delta_{k-s} y^A_{(s)}} = -\frac{d}{dx_1} p_{A/s+1} \qquad \frac{\delta_{k-s} H}{\delta_{k-s} p_{A/s+1}} = \frac{d}{dx_1} y^A_{(s)}, \tag{13}$$

<u>where</u> $0 \leq s \leq k-1$.

<u>Proof</u>: From the Definition (2) it follows

$$\frac{\delta_{k-s} H}{\delta_{k-s} y^A_{(s)}} = p_{A/s} - \frac{\delta_{k-s} L}{\delta_{k-s} y^A_{(s)}},$$

which, according to (11) gives

$$\frac{\delta_{k-s} H}{\delta_{k-s} y^A_{(s)}} = \sum_{0}^{k-s}{}_{\ell} (-1)^{\ell} \frac{d^{\ell}}{dx_1^{\ell}} \frac{\delta_{k-\ell-s} L}{\delta_{k-\ell-s} y^A_{(\ell+s)}} - \frac{\delta_{k-s} L}{\delta_{k-s} y^A_{(s)}}$$

$$= \sum_{1}^{k-1}{}_{\ell} (-1)^{\ell} \frac{d^{\ell}}{dx_1^{\ell}} \frac{\delta_{k-\ell-s} L}{\delta_{k-\ell-s} y^A_{(\ell+s)}}$$

$$= -\frac{d}{dx_1}\Big(\sum_{0}^{k-(s+1)}{}_{\ell} (-1)^{\ell} \frac{d^{\ell}}{dx_1^{\ell}} \frac{\delta_{k-\ell-s} L}{\delta_{k-\ell-s} y^A_{(\ell+s+1)}} \Big).$$

Using again (11) we obtain equations (13). It is also clear that $(\partial h/\partial x_a) = -(\partial \mathfrak{L}/\partial x_a)$ and that in the above expressions $\delta_o H/\delta y^A_{(k)} \equiv 0$ making H independent of $p_{A/0}$.

We have shown that Euler-Lagrange equations imply Hamilton equations. Now, according to equations (13)

$$\frac{\delta H}{\delta y^A} = -\frac{d}{dx_1} \Sigma_{\ell} (-1)^{\ell} \frac{d^{\ell}}{dx_1^{\ell}} \frac{\delta L}{\partial y^A_{(\ell)}} = \Sigma_{\ell} (-1)^{\ell+1} \frac{d^{\ell+1}}{dx_1^{\ell+1}} \frac{\delta L}{\delta y^A_{(\ell)}},$$

where the subscripts have been omitted for sake of simplicity. But

$$\frac{\delta H}{\delta y^A} = p_{A/0} - \frac{\delta L}{\delta y^A} = -\frac{\delta L}{\delta y^A}$$

and so we have

$$\frac{\delta L}{\delta y^A} + \Sigma_{\ell} (-1)^{\ell+1} \frac{d^{\ell+1}}{dx_1^{\ell+1}} \frac{\delta L}{\delta y^A_{(\ell)}} = \Sigma_{\ell} (-1)^{\ell} \frac{d^{\ell}}{dx_1^{\ell}} \frac{\delta L}{\delta y^A_{(\ell)}} = 0,$$

showing the equivalence of both equations. □

REMARK (2). The coordinate x_1 singled out for the Hamiltonian formulation is <u>arbitrarily</u> chosen. If we consider the Minkowsky manifold M_4, then x_1 is not necessarily the t. So, in spite of its apparent non-covariance, equations $(5)_3$ are (relativistic) equivalent to the covariant equations (5). The method is nothing but a more concise way of writing the field equations.

Definition (2) maintains the usual properties of functional derivation. In particular, let us show the following "chain rule":

PROPOSITION (2). Let $F = \int f\, d^{n-1}x$, where $f = f(x_a, y^A, y^A_{a(s)})$ is a smooth function on $\mathbb{R}^K$. Suppose that all $y^A_{(\ell)}$ are functionals of other variables z^B, $z^B_{(i)}$, $1 \le i \le k$ and their higher order partial derivatives with respect to x_a. Then

$$\frac{\delta_r F}{\delta_r z^B_{(i)}} = \Sigma_A \Sigma_\ell \int \frac{\delta_r F}{\delta_r y^A_{(\ell)}} \frac{\delta_r y^A_{(\ell)}}{\delta_r z^B_{(i)}}\, d^{n-1}x \,.$$

Proof: If we perform a variation $\delta_r y^A_{(\ell)}$, we have:

$$\delta_r F = \Sigma_A \Sigma_\ell \int \left(\frac{\partial f}{\partial y^A_{(\ell)}} \delta_r y^A_{(\ell)} + \Sigma_a \frac{\partial f}{\partial y^A_{(\ell)a}} \delta_r y^A_{(\ell)a} + \ldots\right) d^{n-1}x.$$

From a similar procedure on successive partial integrations as developed to obtain the Euler-Lagrange equations (5) (Gelfand & Fomin (1963), p.174) the above expression becomes:

$$\delta_r F = \Sigma_A \Sigma_\ell \int \left(\frac{\delta_r F}{\delta_r y^A_{(\ell)}}\right)(\delta_r y^A_{(\ell)})\, d^{n-1}x.$$

Now, as $y^A_{(\ell)}$ are functionals of $z^B_{(i)}$ and their higher order partial derivatives with respect to x_a, performing a δ_r variation on $z^B_{(i)}$ we have:

$$\delta_r y^A_{(\ell)}(x) = \Sigma_B \Sigma_i \int \left(\frac{\delta_r y^A_{(\ell)}(x)}{\delta_r z^B_{(i)}(\bar{x})}\right)(\delta_r z^B_{(i)}(\bar{x}))\, d^{n-1}\bar{x}.$$

But, if F is also a functional of z then

$$\delta_r F = \Sigma_B \Sigma_i \int \frac{\delta_r F}{\delta_r z^B_{(i)}} \delta_r z^B_{(i)}\, d^{n-1}x.$$

Substituting $\delta_r y^A_{(\ell)}$ into the expression for $\delta_r F$ and comparing the result with the above expression one has the desired "chain rule" for our generalized functional. $\square$

In the definition of the Hamiltonian we have selected the set of coordinates $Y = \{y^A, \ldots, y^A_{(k-1)}\}$ and $P = \{p_{A/1}, \ldots, p_{A/k}\}$ to be the "canonical set of conjugate variables". For this it is inferred that the density $\mathfrak{L}$ is regular with respect to the variables $y^A_{(k)}$. This assures the independence of Y and P in the sense that if they are not then we may assume that the theory is not of order k but of order $r < k$. For example if $\mathfrak{L}$ is given by $\mathfrak{L} = y\square y - m^2 y$, where m is the mass and $\square$ is the D'Alembertian operator, then for a theory of order 2, $p_1 = -y_{(1)}$ and $p_2 = y$ are not independent of y and $y_{(1)}$. This is not a 2nd order theory.

In fact, the higher order momenta are functions of y and their derivatives up to order $2k-s$:

$$p_{A/s} = p_{A/s}\left(y^A, \frac{\partial y^A}{\partial x_1}, \ldots, \frac{\partial^{2k-s} y^A}{\partial x_1^{2k-s}}\right), \qquad 1 \leq s \leq k.$$

This shows that for $s \neq k$, the p's are functions of derivatives greater than k (the order of the Lagrangian) and so (at least theoretically) we need to extend the notion of regularity for the Lagrangian density. This has already been remarked for Generalized Particle Theory. We will return to this point in III.3 for a geometrical discussion.

REMARK (3). We may have a more symmetrical expression for equations (13) if we put $p_{A/-1} \equiv 0$ like it was shown in

Chapter II, §4.2. In the present situation our canonical variables are again $y^A_{(i)}$ and $p_{A/i}$, with Hamiltonian equations being

$$\frac{\delta_{k-i} H}{\delta_{k-i} y^A_{(i)}} = -\frac{d}{dx_1} p_{A/i} \; ; \qquad \frac{\delta_{k-i} H}{\delta_{k-i} p_{A/i}} = \frac{d}{dx_1} y^A_{(i)} \tag{13$_1$}$$

where now $0 \le i \le k-1$. From the definition of the Hamiltonian functional, it is clear that the density h is a function of $(x_a, y^A_{(i)}, p_{A/i})$ and we proceed by analogy with Chapter II to define the corresponding Legendre transformation (we recognize that this procedure obscures the other field variables involved in the theory). Finally, it is also clear that

$$\frac{\delta_{k-s} H}{\delta_{k-s} p_{A/i}} = \frac{\partial h}{\partial p_{A/i}} ,$$

since derivatives of $p_{A/s+1}$ do not appear in h.

2.5. The Generalized Electrodynamics of Bopp-Podolsky.

We have seem above an attempt to extend Jacobi-Ostrogradsky's method for field theories with higher order derivatives. Let us see now an interesting example (Bopp (1940) and Podolsky (1942)) of a generalization of Electromagnetic theory with second order derivatives.

We start with a brief review of Electromagnetism in absence of charge and currents. Maxwell equations are:

$$\left.\begin{aligned} \text{curl } E &= -(1/c)\partial H/\partial t \\ \text{div } H &= 0 \end{aligned}\right\} M_1 \qquad \left.\begin{aligned} \text{Curl } H &= (1/c)\partial E/\partial t \\ \text{div } E &= 0 \end{aligned}\right\} M_2$$

where $E = (E_1, E_2, E_3)$ is the electric field and $H = (H_1, H_2, H_3)$ is the magnetic field. These functions are defined on a Minkowsky manifold with spatial coordinates x_1, x_2, x_3 and scalar $x_y = ict$.

REMARK (4). Maxwell equations can be expressed in terms of differential forms. R. Debever (1951) defined an Electromagnetic space as being M_4 endowed with two differential forms of degree 2, one of which is exact. A more modern treatment of Debever point of view can be founded in the book of Flanders (1963) on differential forms in Physics.

The equations M_1 and M_2 may be put in a tensorial form. For this we introduce the antisymmetric tensor field $F_{\alpha\beta}$ defined by

$$F_{\alpha\beta} = \begin{pmatrix} 0 & H_3 & -H_2 & -iE_1 \\ -H_3 & 0 & H_1 & -iE_2 \\ H_2 & -H_1 & 0 & -iE_3 \\ iE_1 & iE_2 & iE_3 & 0 \end{pmatrix}.$$

M_1 and M_2 assume the form

$$\frac{\partial F_{\alpha\beta}}{\partial x_\gamma} + \frac{\partial F_{\beta\gamma}}{\partial x_\alpha} + \frac{F_{\gamma\alpha}}{\partial x_\beta} = 0 \qquad \bar{M}_1$$

$$\frac{\partial F_{\alpha\beta}}{\partial x_\beta} = 0 \qquad \bar{M}_2$$

The tensor $F_{\alpha\beta}$ can be expressed in terms of the <u>potential</u> of the field $A_\alpha = (A_1, A_2, A_3, iA_o)$ by

$$F_{\alpha\beta} = \frac{\partial A_\beta}{\partial x_\alpha} - \frac{\partial A_\alpha}{\partial x_\beta}$$

(and so $F_{44} \equiv 0$). The potential is introduced to simplify the calculations but it is not uniquely determinated. In fact the addition to A_α of $\partial F/\partial x_\alpha$ for an arbitrary smooth real function F does not alter the field. The Maxwell equations are indeed invariant under any action of type

$$A_\alpha \mapsto A_\alpha + \partial F/\partial x_\alpha .$$

This is called a gauge transformation and in terms of differential forms it means that the difference between the new form and the old is a closed form. Now, to obtain one set of Maxwell equations we need to impose some subsidiary condition, called gauge condition. We choose Lorentz's gauge, that is

$$\partial A_\alpha/\partial x_\alpha = 0.$$

The equations $\bar{M}_2$ can be put in the form,

$$\Box A_\alpha - \frac{\partial}{\partial x_\alpha}\left(\frac{\partial A_\beta}{\partial x_\alpha}\right) = 0.$$

If $\partial A_\beta/\partial x_\alpha \neq 0$ then we make a transformation $\bar{A}_\alpha = A_\alpha + \partial g/\partial x_\alpha$, where g is such that $\Box g = -\partial A_\alpha/\partial x_\alpha$, and then

$$\frac{\partial \bar{A}_\alpha}{\partial x_\alpha} = \frac{\partial A_\alpha}{\partial x_2} + \Box g = 0.$$

Maxwell equations $\bar{M}_2$ can be obtained from the variational approach. Assuming that the Lagrangian density is a function of A_α and $\partial A_\alpha/\partial x_\beta$ one obtains, as usually, the Euler-Lagrange field equations. The Lagrangian density for the Electromagnetic field is

$$\mathfrak{L} = (1/2)(E^2-H^2) = -(1/4)F_{\alpha\beta}F_{\alpha\beta}$$ (summation convention is adopted).

The field equations are

$$\frac{\partial}{\partial x_\alpha}\left(\frac{\partial \mathfrak{L}}{\partial(\partial A_\beta/\partial x_\alpha)}\right) = 0 \qquad (\text{or } \partial\mathfrak{L}/\partial A_\beta = 0).$$

This gives equations $\bar{M}_2$ since $F_{\alpha\beta} = \partial\mathfrak{L}/\partial(\partial A_\beta/\partial x_\alpha)$. Now, if we develop the derivative $d\mathfrak{L}/dx_\alpha$, (recall that $\mathfrak{L} = \mathfrak{L}(A_\beta, \partial A_\beta/\partial x_\gamma))$ and if we take into account the field equations, we obtain

$$\frac{\partial}{\partial x_\alpha}\left(\mathfrak{L}\,\delta_\alpha^\beta - \frac{\partial \mathfrak{L}}{\partial(\partial A_\gamma/\partial x_\alpha)}\frac{\partial A_\gamma}{\partial x_\beta}\right) = 0.$$

The term in the interior of the bracket is the <u>energy momentum tensor</u> $T_{\alpha\beta}$. It is called so because we can define a conserved quantity

$$P_\alpha = -i\int T_{\alpha 4}\, dx_1 dx_2 dx_3$$

for which the spatial components (1,2,3) represent the momentum of the field and its time component ($\div$ i) is the Energy of the field.

The Hamiltonian formulation for the Electromagnetic field is developed as usual but it has an important feature which must be stressed: there is no canonical momentum conjugate to $A_4 = iA_o$. In fact, putting $\mathfrak{H} = \pi_\alpha A_\alpha - \mathfrak{L}$, where $\pi_\alpha = \partial\mathfrak{L}/\partial(\partial A_\alpha/\partial x_4)$, from the fact that $\pi_\alpha = F_{\alpha 4}$, we see that $\pi_4 \equiv 0$, since $F_{44} \equiv 0$.

Let us now see the <u>Bopp-Podolsky approach</u>. The usual conditions are assumed: (i) the field equations should be linear, (ii) they are derived from variational pinciples, (iii) the Lagrangian density is a sum of densities representing the fields, the particles and their interactions (but we assume, for simplicity, that there is no particles and interaction terms). The only new hypothesis is that the density is a function of type

$$\mathfrak{L} = \mathfrak{L}(y^\alpha, y^\alpha_\beta, y^\alpha_{\beta\gamma}),$$

that is, $\mathfrak{L}$ is a function of the potential and their first and second derivatives with respect to the Minkowsky variables:

$$y^\alpha = A_\alpha = (A_1, A_2, A_3, iA_o)$$

$$y^\alpha_\beta = \partial A_\alpha/\partial x_\beta = A_{\alpha,\beta}\ ; \qquad y^\alpha_{\beta\gamma} = \partial^2 A_\alpha/\partial x_\beta \partial x_\gamma = A_{\alpha,\beta\gamma}\ .$$

The hamiltonization of the theory is obtained by the introduction of new coordinates:

$$q_\alpha = y^\alpha, \qquad Q_\alpha = y^\alpha_4 = \partial A_\alpha/\partial t = A_{\alpha,4}\ ,$$

(in the following Greek symbols run from 2 to 4 and the Latin from 1 to 3). The momenta are defined by

$$p_{\alpha/1} = \frac{\partial \mathfrak{L}}{\partial \dot{q}_\alpha} - \frac{\partial}{\partial t}\left(\frac{\partial \mathfrak{L}}{\partial \ddot{q}_\alpha}\right) - \frac{\partial}{\partial x_j}\left(\frac{\partial \mathfrak{L}}{\partial \dot{q}_{\alpha,j}}\right)$$

$$p_{\alpha/2} = \partial \mathfrak{L}/\partial \ddot{q}_\alpha$$

where the dots are the derivatives with respect to the time (i.e. with respect to the variable x_4). We remark that

$$\dot{q}_{\alpha,j} = \frac{\partial \dot{q}_\alpha}{\partial x_j} = \frac{\partial}{\partial x_j}\left(\frac{\partial A_\alpha}{\partial t}\right) = \frac{\partial}{\partial x_j} A_{\alpha,4} = icA_{\alpha,4j} = icA_{\alpha,j4}$$

and so we may re-write $p_{\alpha/1}$ in the form

$$p_{\alpha/2} = \frac{1}{ic}\left[\frac{\partial \mathfrak{L}}{\partial A_{\alpha,4}} - \frac{\partial}{\partial x_4}\frac{\partial L}{\partial A_{\alpha,44}} - \frac{\partial}{\partial x_j}\left(\frac{\partial \mathfrak{L}}{\partial A_{\alpha,4j}} + \frac{\partial \mathfrak{L}}{\partial A_{\alpha,j4}}\right)\right]$$

$$= \frac{1}{ic}\left[\frac{\partial \mathfrak{L}}{\partial y^\alpha_4} - \frac{\partial}{\partial x_4}\frac{\partial \mathfrak{L}}{\partial y^\alpha_{44}} - \frac{\partial}{\partial x_j}\left(\frac{\partial \mathfrak{L}}{\partial y^\alpha_{4j}} + \frac{\partial \mathfrak{L}}{\partial y^\alpha_{j4}}\right)\right].$$

The Hamiltonian density is defined by

$$\mathfrak{H} = p_{\alpha/1}\,\dot{q}_\alpha + p_{\alpha/2}\,Q_\alpha - \mathfrak{L}$$

and we suppose that $\mathcal{H}$ is a function of type

$$\mathcal{H} = \mathcal{H}(q_\alpha, q_{\alpha,ij}, Q_{\alpha,i}, p_{\alpha/1}, p_{\alpha/2}).$$

The energy is defined by

$$H = \int \mathcal{H}\, dx_1 dx_2 dx_3$$

and from the definition of s - functional derivatives we obtain the canonical equations

$$\frac{\delta_2 H}{\delta_2 p_{\alpha/1}} = \dot{q}_\alpha \qquad \frac{\delta_2 H}{\delta_2 q_2} = -\dot{p}_{\alpha/1}$$

$$\frac{\delta_1 H}{\delta_1 p_{\alpha/2}} = \dot{Q}_\alpha \qquad \frac{\delta_1 H}{\delta_1 Q_\alpha} = -\dot{p}_{\alpha/2} .$$

The definition of the energy-momentum tensor is

$$ict_{\alpha\beta} = \mathfrak{L}\, \delta_\alpha^\beta - y_\alpha^u p_{u\beta} - y_{\alpha\lambda}^u p_{u\lambda\beta}$$

where

$$p_{u\beta} = \frac{\partial \mathfrak{L}}{\partial y_\beta^u} - \frac{\partial}{\partial x_\alpha}\left(\frac{\partial \mathfrak{L}}{\partial y_{\alpha\beta}^u}\right)$$

$$p_{u\lambda\beta} = \frac{\partial \mathfrak{L}}{\partial y_{\lambda\beta}^u} .$$

For the above formalism we will consider the following density:

$$\mathfrak{L} = -(1/2)[(1/2)F_{\alpha\beta}F_{\alpha\beta} + a^2(\partial F_{\alpha\beta}/\partial x_\beta)^2]$$

where a is a constant. The Euler-Lagrange field equations are:

$$(1-a^2\square)F_{\alpha u,u} = 0$$

and the energy-momentum tensor is obtained from the values

$$\partial \mathfrak{L}/\partial g^{\alpha}_{\nu} = \partial \mathfrak{L}/\partial A_{\alpha,\nu} = F_{\alpha,\nu}$$

$$\partial \mathfrak{L}/\partial y^{\alpha}_{\nu\lambda} = \partial \mathfrak{L}/\partial A_{\alpha,\nu\lambda} = -a^2(F_{\nu\beta,\beta}\delta_{\lambda\alpha} - F_{\alpha\beta,\beta}\delta_{\nu\lambda}).$$

Hence

$$ict_{\alpha\beta} = \mathfrak{L}\,\delta_{\alpha\beta} - A_{u,\alpha}(1-a^2\Box)F_{u\beta} + a^2F_{u\beta,\alpha}F_{u\nu,\nu}\,.$$

The Hamiltonian density is

$$\mathfrak{H} = -ict_{44} = \mathfrak{L} - A_{\alpha,4}(1-a^2\Box)F_{\alpha 4} + a^2F_{\alpha 4,4}F_{\alpha\beta,\beta}\,,$$

and the momenta

$$p_{4/1} = (a^2/ic)F_{j4,j4}\,; \quad p_{j/1} = (1/ic)(1-a^2\Box)F_{j4}$$

$$p_{4/2} = 0 \qquad ; \quad p_{j/2} = -(a^2/c^2)F_{j\beta,\beta}.$$

Contrary to the usual situation we see that $p_{4/1} \neq 0$ but $p_{4/2}$ maintains the same problem. Podolsky suggests the following modified Lagrangian:

$$\mathfrak{L} = -(1/2)(A_{\alpha,\beta}A_{\alpha,\beta} + a^2A_{\alpha,\beta\beta}A_{\alpha,\nu\nu}),$$

which gives, from a direct calculation, the form:

$$\mathfrak{L} = -(1/2)[(1/2)F_{\alpha\beta}F_{\alpha\beta}+a^2F_{\alpha\beta,\beta}F_{\alpha\beta,\beta}] - (1/2)A_{\alpha,\alpha}(1-a^2\Box)A_{\beta,\beta}+V_{\beta,\beta}$$

where

$$V_{\beta} = (1/2)(A_{\alpha,\alpha}A_{\beta}-A_{\alpha}A_{\beta,\alpha}) + (a^2/2)A_{\alpha,\alpha}(A_{\nu,\nu\beta}-2A_{\beta,\nu\nu}).$$

The comparison with the original Lagrangian density shows that the first term in the above Lagrangian is the former Lagrangian. V_{β} is a divergent function and the intermediate term is zero when we consider the Lorentz condition. This shows the equivalence of this Lagrangian with the original ones.

The advantage of such new density is the non-nullity of the momenta:

$$p_{\alpha/1} = \frac{1}{c^2}\,(1-a^2\Box)\,\frac{\partial A_\alpha}{\partial t}$$

$$p_{\alpha/2} = \frac{1}{c^2}\,a^2\Box A_\alpha\,.$$

The Hamiltonian density is

$$\mathcal{H} = -\mathfrak{L} - A_{\alpha,4}(1-a^2\Box)A_{\alpha,4} - a^2 A_{\alpha,44}\Box A_\alpha\,.$$

As Maxwell's theory is very successful in describing electromagnetic phenomena, the reader may be asking himself about the reasons to introduce a Lagrangian density which takes to modified Maxwell's equations. As a matter of fact these modifications are introduced as an attempt to get rid of the divergences that appear when the quantization is made departing from the usual first order theory. The subject will be further discussed later in III.4.

2.6. The de Donder-Hamilton field equations.

In the theory hitherto developed we have singled out one variable, x_1, in the Euclidean space $\mathbb{R}^n$. We shall now develop a covariant formulation; convariance means that equations that describe physical systems must keeps the same form when a Lorentz transformation is performed. This can also be stated as: "the laws of physics are the same in any inertial reference system". The necessity of covariance imposes that space and time be treated on the same footing. We consider the x's fixed. From (7) we put

$$p_{A/a(\ell)} = \delta\mathfrak{L}/\delta y^A_{a(\ell)}\,, \tag{14}$$

or explicitly,

$$p_{A/a(\ell)} = p_{A/a_1\dots a_\ell} = \sum_{0}^{k-\ell}{}_j (-1)^j \frac{d^j}{dx_{b_1}\dots dx_{b_j}}\left(\frac{\partial \mathcal{L}}{\partial y^A_{a_1\dots a_\ell b_1\dots b_j}}\right), \quad (14)_1$$

The equations of the extremals in the canonical form are obtained identically to the standard case if we define the density by

$$h = p_{A/a(r+1)}\, y^A_{a(r+1)} - \mathcal{L}, \qquad 0 \leq r \leq k-1.$$

These equations are:

$$\frac{\partial h}{\partial p_{A/a(r+1)}} = \frac{d}{dx_{a_{r+1}}} y^A_{a(r)}\ ; \quad \frac{\partial h}{\partial y^A_{a(r)}} = -\frac{d}{dx_{a_{r+1}}} p_{A/a(r+1)}, \quad (15)$$

called de Donder-Hamilton field equations (cf. de Donder (1935)). Of course equality (14) or $(14)_1$ is the definition of higher order momentum for the present formalism.

If we want to consider the general situation, that is, the variation of the action integral is total, then we include in the theory a generalization of the energy-momentum tensor. This treatment was proposed by Barut and Mullen (1962)(b), Borneas (1969) and more recently by Musicki (1978)(b) and is obtained when the Euler-Lagrange field equations are satisfied. In this situation the first term of (9) in §2.3 disappears and making the corresponding substitutions we found

$$\int_D \sum_{0}^{k-1}{}_r \frac{d}{dx_{a_{r+1}}}\left[p_{A/a(r+1)}\, \delta y^A_{a(r)} - T_b^{a_{r+1}}\, \delta x_b\right] d^n x,$$

where

$$T_b^{a_{r+1}} \overset{\text{def}}{=} p_{A/a(r+1)}\, y^A_{a(r)b} - \delta_b^{a_{r+1}}\, \mathcal{L},$$

is the higher-order energy-momentum tensor, with δ_i^j being the Kroenecker delta.

2.7. Canonical transformations and Poisson brackets.

We begin this paragraph with the formalism proposed in §2.4. Consider a transformation of the canonical variables of the type:

$$Y^B_{(r)} = Y^B_{(r)}(y^A_{(s)}, p_{A/s+1}); \quad P_{B/r+1} = P_{B/r+1}(y^A_{(s)}, p_{A/s+1}), \quad (16)$$

with $1 \leq A,B \leq m$, $0 \leq r,s \leq k-1$, from a point of coordinates (y,p) to another point (Y,P). From the definition of the Hamiltonian density (12) we use the fact that the corresponding Hamilton equations are the generalized Euler-Lagrange equations of the functional

$$\Sigma_A \Sigma_s \int (p_{A/(s+1)} dy^A_{(s)} - h\ dx_1) d^{n-1}x.$$

Therefore for the new Hamiltonian K in the coordinates (Y,P), the transformation (15) will be canonical if and only if

$$\Sigma_A \Sigma_s \int [(p_{A/s+1} dy^A_{(s)} - h\ dx_1) - (P_{A/s+1} dY^A_{(s)} - k\ dx_1)] d^{n-1}x = dG ,$$

where G is the generating function of the above transformation and k the corresponding density for K. As it is well known, there are four forms for G: $G_1(y,Y)$, $G_2(y,P)$, $G_3(p,Y)$, $G_4(p,P)$. Let us take, for example G_1 and G_2. We obtain the following set of equations (<u>omitting again the subscripts for the variational derivative</u>):

$$p_{A/s+1} = \frac{\delta G_1}{\delta y^A_{(s)}}; \qquad P_{A/s+1} = -\frac{\delta G_1}{\delta Y^A_{(s)}} \quad \text{and} \quad K = H + \frac{\partial G_1}{\partial x_2}$$

and for G_2

$$p_{A/s+1} = \frac{\delta G_2}{\delta y^A_{(s)}}; \qquad Y^A_{(s)} = \frac{\delta G_2}{\delta P_{A/s+1}} \quad \text{and} \quad K = H + \frac{\partial G_2}{\partial x_1}.$$

In particular, let us consider

$$G_2 = \Sigma_A \Sigma_s \int y^A_{(s)} P_{A/s+1} \, d^{n-1}x + \epsilon K(x, y^A_{(0)}, P_{A/s+1})$$

(where ϵ is small). Then

$$Y^A_{(s)} = \frac{\delta G_2}{\delta P_{A/s+1}} = y^A_{(s)} + \epsilon \frac{\delta K}{\delta P_{A/s+1}}$$

$$p_{A/s+1} = \frac{\delta G_2}{\delta y^A_{(s)}} = P_{A/s+1} + \epsilon \frac{\delta K}{\delta y^A_{(s)}} .$$

So, if we consider the <u>infinitesimal canonical transformation</u>

$$P_{A/s+1} - p_{A/s+1} = \delta p_{A/s+1} ; \qquad Y^A_{(s)} - y^A_{(s)} = \delta y^A_{(s)} ,$$

we obtain

$$\delta y^A_{(s)} = \epsilon \frac{\delta K}{\delta P_{A/s+1}} ; \qquad \delta p_{A/s+1} = -\epsilon \frac{\delta K}{\delta y^A_{(s)}}$$

and so

$$\delta y^A_{(s)} = dy^A_{(s)} ; \qquad \delta p_{A/s+1} = dp_{A/s+1}$$

when $\epsilon = dx_1$, $K = H$, (where $P_{A/s+1}$ was replaced by $p_{A/s+1}$ since they differ by an infinitesimal term).

DEFINITION (3). Let F and G be two functionals of the canonical variables $y^A_{(s)}$ and $p_{A/s+1}$. Then the <u>Poisson brackets</u> $[F,G]$ is

$$[F,G]_{(y,p)} = \Sigma_A \Sigma_s \int \left(\frac{\delta F}{\delta y^A_{(s)}} \frac{\delta G}{\delta p_{A/s+1}} - \frac{\delta F}{\delta p_{A/s+1}} \frac{\delta G}{\delta y^A_{(s)}} \right) d^{n-1}x$$

and the <u>Lagrange brackets</u> is

$$\{f,g\}_{(y,p)} = \Sigma_A \Sigma_s \int \left(\frac{\delta y^A_{(s)}}{\delta f} \frac{\delta p_{A/s+1}}{\delta g} - \frac{\delta y^A_{(s)}}{\delta g} \frac{\delta p_{A/s+1}}{\delta f} \right) d^{n-1}x.$$

Some of its main properties are:

$$\text{(i)} \quad \frac{d}{dx_1} y^A_{(s)} = [y^A_{(s)}, H] ; \quad - \frac{d}{dx_1} p_{A/s+1} = [H, p_{A/s+1}]$$

$$\text{(ii)} \quad [y^A_{(s)}, y^B_{(r)}] = [p_{A/s+1}, p_{B/r+1}] = 0$$

$$\text{(iii)} \quad [y^A_{(s)}(x), p_{B/r+1}(\bar{x})] = \delta^A_B \, \delta^s_{r-1} \, \delta(x-\bar{x})$$

$$\text{(iv)} \quad \frac{dF}{dx_1} = \frac{\delta F}{\delta x_1} + [F, H]$$

$$\text{(v)} \quad \{y^A_{(s)}, y^B_{(r)}\} = \{p_{A/s+1}, p_{B/r+1}\} = 0$$

$$\text{(vi)} \quad \{y^A_{(s)}(x), p_{B/r+1}(\bar{x})\} = \delta^A_B \, \delta^s_{r-1} \, \delta(x-\bar{x}),$$

which maintains the analogy with the standard theory.

Suppose now that transformation (16) is, at least, a local diffeomorphism and x_1-independent. <u>For the new Hamiltonian</u>

$$K(Y^A_{(s)}, P_{A/s+1}) = H(y^A_{(s)}(Y^B_{(r)}, P_{B/r+1}), p_{A/s+1}(Y^B_{(r)}, P_{B/r+1})),$$

(15) <u>is canonical if and only if the Hamiltonian equations</u> have the form:

$$- \frac{d}{dx_1} P_{A/s+1} = \frac{\delta K}{\delta Y^A_{(s)}} ; \quad \frac{d}{dx_1} Y^A_{(s)} = \frac{\delta K}{\delta P_{A/s+1}} . \tag{17}$$

From the Hamiltonian equations (13) and the "chain rule" we have the following equalities:

$$\left.\begin{aligned} &\Sigma_B \Sigma_r \int \left(\frac{\delta K}{\delta Y^B_{(r)}} \frac{\delta Y^B_{(r)}}{\delta y^A_{(s)}} + \frac{\delta K}{\delta P_{B/r+1}} \frac{\delta P_{B/r+1}}{\delta y^A_{(s)}} \right) d^n x \\ &= -\Sigma_B \Sigma_r \int \left(\frac{\delta p_{A/s+1}}{\delta Y^B_{(r)}} \frac{dY^B_{(r)}}{dx_1} + \frac{\delta p_{A/s+1}}{\delta P_{B/r+1}} \frac{dP_{B/r+1}}{dx_1} \right) d^n x \end{aligned}\right\} \quad (18)_1$$

$$\left.\begin{aligned} &\Sigma_B \Sigma_r \int \left(\frac{\delta K}{\delta Y^B_{(r)}} \frac{\delta Y^B_{(r)}}{\delta p_{A/s+1}} + \frac{\delta K}{\delta P_{B/r+1}} \frac{\delta P_{B/r+1}}{\delta p_{A/s+1}} \right) d^n x \\ = \; &\Sigma_B \Sigma_r \int \left(\frac{\delta y^A_{(s)}}{\delta Y^B_{(r)}} \frac{dY^B_{(r)}}{dx_1} + \frac{\delta y^A_{(s)}}{\delta P_{B/r+1}} \frac{dP_{B/r+1}}{dx_2} \right) d^n x \end{aligned} \right\} \quad (18)_2$$

Substituting (17) into $(18)_1$ and $(18)_2$ and then equating the coefficients of the field variables Y and P, we obtain the following <u>canonical conditions</u>

$$\left.\begin{aligned} &\frac{\delta Y^B_{(r)}}{\delta y^A_{(s)}} = \frac{\delta p_{A/s+1}}{\delta P_{B/r+1}} \; ; \qquad \frac{\delta Y^B_{(r)}}{\delta p_{A/s+1}} = - \frac{\delta y^A_{(s)}}{\delta P_{B/r+1}} \\ &\frac{\delta P_{B/r+1}}{\delta y^A_{(s)}} = - \frac{\delta p_{A/s+1}}{\delta Y^B_{(r)}} \; ; \qquad \frac{\delta P_{B/r+1}}{\delta p_{A/s+1}} = \frac{\delta y^A_{(s)}}{\delta Y^B_{(r)}} \end{aligned} \right\} \quad (19)$$

PROPOSITION (3) (Invariance with respect to the canonical transformations). <u>The Poisson (resp. Lagrange) brackets of any functional of the field variables of higher order are invariant for any canonical transformation.</u>

<u>Proof</u>: This result is easily obtained from a straightforward calculation. If F and G are two functionals, taking Definition (3), developing $\delta F/\delta y^A_{(s)}$ and $\delta F/\delta p_{A/s+1}$ by the "chain rule" and, finally, substituting the canonical conditions (19) we obtain the desired result, i.e.,

$$[F,G]_{(y,p)} = [F,G]_{(Y,P)} \; . \qquad \square$$

Let us now see the situation described in §2.6. We follow Musicki (1978)(a), considering now the boundary formula

$$\int \sum_{0}^{k-1} r\,(p_{A/a(r+1)}\delta y^A_{a(r)} - h\, dx_{a_{r+1}})\, d\omega_{a_{r+1}}$$ (ω corresponds to a surface domain of integration)

which is equivalent to the de Donder-Hamilton canonical equations (15). So, if we perform a transformation

$$T:\ (y^A_{a(r)}, p_{A/a(r+1)}) \mapsto (Y^A_{a(r)}, P_{A/a(r+1)})$$

with $y^A_{a(0)} = y^A$, etc. and $0 \le r \le k-1$, then T will be canonical if and only if

$$\int (p_{A/a(r+1)}\delta y^A_{a(r)} - h\,\delta x_{a_{r+1}})\, d\omega_{a_{r+1}} =$$

$$= \delta G + \int (P_{A/a(r+1)}\delta Y^A_{a(r)} - \bar{h}\,\delta x_{a_{r+1}})\, d\omega_{a_{r+1}} \tag{20}$$

where $\bar{h}$ is the new Hamiltonian. We proceed as usual; if we take a generating function $G_1 = G_1(y^A_{a(r)}, P_{A/a(r+1)})$ of type 1, defined now by

$$G_1 = G + \int P_{A/a(r+1)}\, Y^A_{a(r)}\, d\omega_{a_{r+1}},$$

then (20) takes the form

$$\int (p_{A/a(r+1)}\delta y^A_{a(r)} - P_{A/a(r+1)}\delta Y^A_{a(r)})\, d\omega_{a_{r+1}} + \delta \int P_{A/a(r+1)} Y^A_{a(r)}\, d\omega_{a_{r+1}}$$

$$+ \int (\bar{h} - h)\,\delta x_{a_{r+1}}\, d\omega_{a_{r+1}} = \delta G_1 .$$

Now, if we put $n^{a_{r+1}}$ for the components of the unit outward normal to the boundary, integrating and comparing with the corresponding coefficients yields (where now δ means the usual functional derivative as defined in (7)):

$$p_{A/a(r+1)} = n^{a_{r+1}} \frac{\delta G_1}{\delta y^A_{a(r)}} ; \qquad Y^A_{a(r)} = n^{a_{r+1}} \frac{\delta G_1}{\delta P_{A/a(r+1)}}$$

$$\frac{\delta G_1}{\delta x_{a_{r+1}}} = \bar{K}-K, \quad \text{where} \quad \bar{K} = \int \bar{h}\, d\omega_{a_{r+1}} , \qquad K = \int h\, d\omega_{a_{r+1}} .$$

Similarly, for a function of type $G_2 = G_2(p_{A/s(r+1)}, Y^A_{a(r)})$

$$y^A_{a(r)} = -n^{a_{r+1}} \frac{\delta G_2}{\delta p_{A/a(r+1)}} ; \qquad P_{A/a(r+1)} = -n^{a_{r+1}} \frac{\delta G_2}{\delta Y^A_{a(r)}}$$

$$\frac{\delta G_2}{\delta x_{a_{r+1}}} = \bar{K} - K.$$

The above equations are a system of functional differential equations and solving this system we obtain new canonical variables as functionals of the old ones. The Poisson brackets is defined in the present situation by

$$[F,G] = \int \left(\frac{\delta F}{\delta y^A_{a(r)}} \frac{\delta G}{\delta p_{A/a(r+1)}} - \frac{\delta F}{\delta p_{A/a(r+1)}} \frac{\delta G}{\delta y^A_{a(r)}}\right) d\omega_{a_{r+1}} ,$$

(we leave to the reader the corresponding definition for the Lagrange brackets).

III.3. - The formalism on Jet manifolds

3.1. Preliminaries.

In Chapter I we have seen some aspects of Jet theory; in the forthcoming paragraphs we will be concerned with their utility.

Jet manifolds can be constructed from smooth mappings between differentiable manifolds and from sections of a given fibered manifold. For our convenience, we shall consider only trivial fibered manifolds with a bundle structure, i.e., if (M,p,N) is a fibered manifold then $M = N\times E$, where the fiber E is a m-dimensional manifold. Since (M,p,N) is trivial we may identify smooth mappings from N to E with sections of (M,p,N) as well as their k-jets. We suppose also that N is a n-dimensional oriented manifold with volume form $\omega = dx_1 \wedge \ldots \wedge dx_n$.

Let (M,p,N) be as above and (J^kM,α^k,N) the corresponding k-jet extension. As (M,p,N) is supposed to be trivial it is possible to show that the same is true for (J^kM,α^k,N). For the sake of brevity we write J^kM for the triple (J^kM,α^k,N); we recall that $\alpha^k\colon J^kM \to N$ is the source projection. Any point of J^kM will be represented as

$$(x_a, y^A, z^A_{a(i)}), \qquad 1 \le i \le k$$

where $x_a = (x_1,\ldots,x_n)$; $y^A = (y^1,\ldots,y^m)$ are local coordinates for M such that x_a are the corresponding coordinates for N induced by the projection p and

$$z^A_{a(i)} = z^A_{a_1\ldots a_i}, \qquad 1 \le i \le r, \quad 1 \le a_1 \le \ldots \le a_i$$

are functions of x_a.

Let $u \in \mathrm{Sec}(J^kM)$. If there is a section $s \in \mathrm{Sec}(M)$ such that $u = \tilde{s}^k$ then u is said k-jet prolongation (or extension) of s. In this case u is locally given in a point $x \in N$ by $(x_a, y^A, y^A_{a(i)})$ where now we put

$$z^A_{a(i)}(u(x)) = \frac{\partial^i (y^A \circ s)}{\partial x_{a_1} \cdots \partial x_{a_i}}(x) = y^A_{a(i)} \, .$$

DEFINITION (1). Let $G\colon J^kM \to \mathbb{R}$ be a smooth function defined on an open set V. The formal derivative of G with respect to the variable x_a is the function

$$d_aG\colon (\rho^{k+1}_k)^{-1}(V) \subset J^{k+1}M \to \mathbb{R}$$

defined by

$$d_aG(\tilde{u}^1(\bar{x})) = (\partial(G \circ u)/\partial \bar{x})_{\bar{x}} \tag{1}$$

where $u \in \mathrm{Sec}(J^kM)$ is such that $\tilde{u}^1(\bar{x}) \in (\rho^{k+1}_k)^{-1}(V) \subset J^{k+1}M$ and $\rho^{k+1}_k\colon J^{k+1}M \to J^kM$ is the projection $\tilde{s}^{k+1}(x) \to \tilde{s}^k$.

In local coordinates we have

$$d_aG = \frac{\partial G}{\partial x_a} + \frac{\partial G}{\partial y^A} z^A_a + \cdots + \frac{\partial G}{\partial z^A_{a(k)}} z^A_{a(k)a} \tag{1$_1$}$$

and along jet prolongations,

$$d_aG = \frac{\partial G}{\partial x_a} + \frac{\partial G}{\partial y^A}\frac{\partial y^A}{\partial x_a} + \cdots + \frac{\partial G}{\partial y^A_{a(k)}} \frac{\partial y^A_{a(k)}}{\partial x_a} = \frac{dG}{dx_a} \tag{1$_2$}$$

i.e., the total derivative of G with respect to x_a. It is clear that

$$d_a(G+F) = d_aG + d_aF; \qquad d_a(G \cdot F) = Gd_aF + Fd_aG,$$

and for the coordinate function y^A, $z^A_{a(i)}$

$$d_ay^A = \frac{\partial y^A}{\partial x_a}; \qquad d_az^A_{a(i)} = \frac{\partial z^A_{a(i)}}{\partial x_a}, \qquad 1 \le i \le k. \tag{2}$$

Let us now recall some facts about variational principles on manifolds, (for more details see, for example, the books of Griffiths (1983) and Hermann (1968)). In the following, N will be a fixed, oriented, connected and compact manifold. Q is an arbitrary manifold and $\pi: Q \to N$ a fibered manifold.

Let I be a differential ideal in the exterior algebra of all differential forms on Q. I is such that $dI \subset I$, that is, closed under exterior differential d. If ω is a n-form on Q then the couple (I,ω) is called an exterior differential system. A submanifold P of Q is an integral manifold of (I,ω) if for every form $v \in I$, $v/P = 0$ and $\omega/P \neq 0$. More generally we say that a mapping $f: P \to Q$ is an integral manifold of (I,ω) if it is a regular immersion $f: P \to Q$ such that $f^*(I) = 0$ and $f^*\omega \neq 0$ (in the following the notations $f^*v = 0$ or $v/P = 0$ will be equally employed).

EXAMPLE (1). Let $Q = J^k(\mathbb{R},E)$, where E is a m-dimensional manifold. A local coordinate system for Q is given by $(t,y^A,z^A_{(1)},\ldots,z^A_{(k)})$, where (t,y^A) is a coordinate system for $\mathbb{R}\times E$. Take N as a closed interval of $\mathbb{R}$ and put $\omega = dt$. The exterior differential system is given by (I,dt), where I is the ideal $\{\theta^A,\theta^A_{(1)},\ldots,\theta^A_{(k-1)},d\theta^A,\ldots,d\theta^A_{(k-1)}\}$ with

$$\theta^A = dy^A - z^A_{(1)}dt,\ldots,\theta^A_{(s-1)} = dz^A_{(s-1)} - z^A_{(s)}dt,$$

where $1 \leq s \leq m$, $2 \leq s \leq k$.

If N is an integral manifold of (I,dt), i.e., $f: N \to J^k(\mathbb{R},E)$ is an integral manifold, then $f^*(\theta^A) = f^*(\theta^A_{(1)}) =\ldots= 0$ (or $\theta^A = \theta^A_{(1)} =\ldots= 0$ along N). In fact, as we will see below, the integral manifolds of $J^k(\mathbb{R},E)$ for the above exterior differential system (I,dt) are the k-jet prolongations of curves of E, $t \mapsto y^A(t)$, since $(dy^A/dt) = z^A_{(1)},\ldots,dz^A_{(k-1)}/dt = d^k y^A/dt^k = z^A_{a(k)}$. □

DEFINITION (2). Let σ be a p-form on the fibered manifold Q. A <u>variational problem</u> with respect to σ on a domain N is given by a mapping $J\colon \mathrm{Sec}(Q) \to \mathbb{R}$ defined by

$$J(u) = \int_N u^*\sigma \qquad \left(\text{or } J(u) = \int_{u(N)} \sigma\right), \tag{3}$$

called <u>Hamilton functional</u> of σ for the sections of $\mathrm{Sec}(Q)$ defined on N (of course we may restrict the integrals to a subdomain $A \subset N$). A <u>variation</u> of (3) is given by a smooth one-parameter family of sections $u_t \in \mathrm{Sec}(Q)$, with $u_o = u$, such that

$$J(u_t) = \int_N u_t^*\sigma.$$

Let X be the tangent vector field along u for u_t at $t = 0$ so that $X(x) \in \mathrm{Ver}\, Q$ at $u(x)$ is the vector tangent to the curve $u_t(x)$:

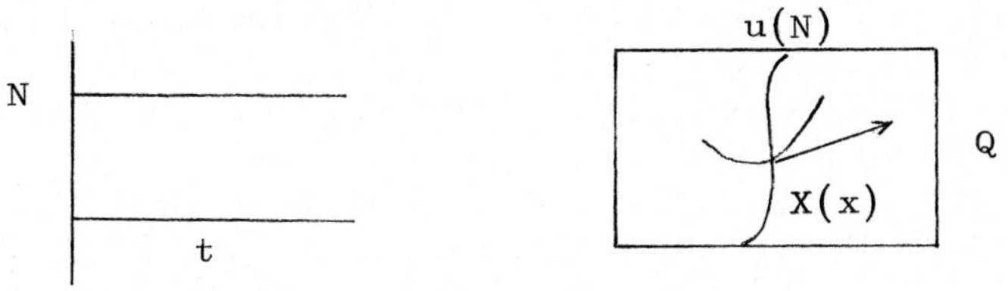

The following formula is well known (see Godbillon (1969), for example):

$$d/dt(u_t^*\sigma)\Big|_{t=0} = u^*[i_X d\sigma + d i_X \sigma] = \mathcal{L}_X u^*\sigma,$$

where $\mathcal{L}$ is the Lie derivative and X is the infinitesimal generator of u_t.

DEFINITION (3) (Hamilton's principle). We say that $u \in Sec(Q)$ is an <u>extremal</u> (or critical section) of (3) if

$$(d/dt)(J(u_t))\Big|_{t=0} = \int_N \mathcal{L}_X u^*\sigma = \delta_u J(X) = 0. \qquad (4)$$

If we restrict our study to the case where (4) vanishes for all variation u_t of u such that $u_t = u$ along ∂N and $X = 0$ along ∂N, then, from Stoke's Theorem, we have that u is an extremal of (3) if and only if

$$\int_N u^*(i_X d\sigma) = 0 \quad \text{(or, if we wish, } \int_{u(N)} i_X d\sigma = 0). \qquad (5)$$

Since this condition must hold for all (vertical) vector fields X along u such that $X|\partial N = 0$ one obtains that along N,

$$u^*(i_X d\sigma) = 0 \quad (\text{or } i(X)d\sigma = 0 \text{ along } u(N)). \qquad (6)$$

Therefore the differential exterior equation (6) gives an equivalent condition for the extremals of the variational problem (3). In the following we will say that $u \in Sec(Q)$ is an extremal of the integral (3) if (6) holds.

REMARK (4). The vector field X tangent to the curve $t \to u_t$ is defined by

$$X(x) = dF(t,x)/dt\Big|_{t=0} \in T_{u(x)}Q,$$

where $F(t,x) = u_t(x)$. Locally u is an embedding, hence we can associate (locally) to X a vector field which is an element of the normal bundle $u^*(TQ/T_u(TN))$ of $u(N)$ in Q, (for further details see Griffiths (1983)).

3.2. The canonical (or structure) one-form on J^kM.

Let (M,p,N) be as usual, where N is n-dimensional and M is $m+n$ dimensional. We start this paragraph by a geometric construction of a system of Pfaffians on J^kM which has an important role in the theory. For this, let us consider the figure which illustrates the situation in question:

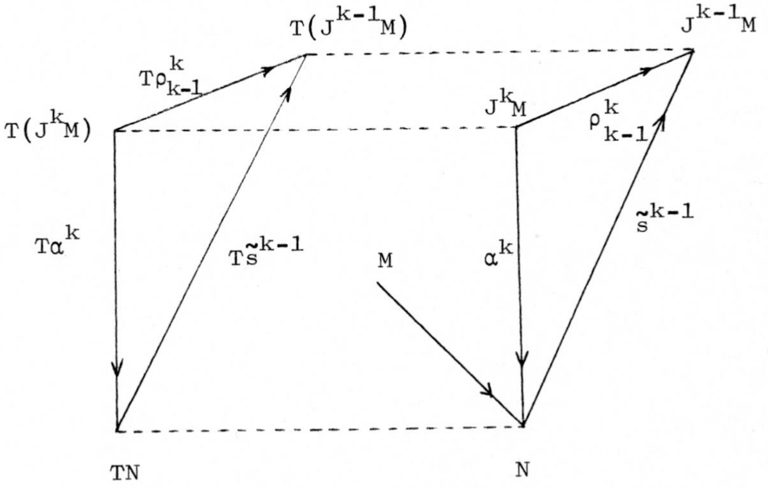

$$\boxed{T\rho^k_{k-1} - T\tilde{s}^{k-1} \circ T\alpha^k : T(J^kM) \to T(J^{k-1}M)} \qquad (*)$$

That is: let z be a point in J^kM, $y = \rho^k_{k-1}(z)$ and $s \in \mathrm{Sec}(M)$ such that $\tilde{s}^{k-1}(x) = y$, where $x = \alpha^k(z)$. Taking the tangent spaces T_xN, $T_y(J^{k-1}M)$ and $T_z(J^kM)$, the corresponding tangent mappings of $\tilde{s}^{k-1}$ and the projections α^k, ρ^k_{k-1} we define the following linear difference

$$T\rho^k_{k-1} - T\tilde{s}^{k-1} \circ T\alpha^k, \qquad (7)$$

giving a linear mapping from $T_z(J^kM)$ to $T_y(J^{k-1}M)$. If $X \in T_z(J^kM)$, then a direct calculation shows that

$$(T\rho^k_{k-1} - T\tilde{s}^{k-1} \circ T\alpha^k)(X) \in Ver_y(J^{k-1}M),$$

where $Ver_y(J^{k-1}M)$ denotes the vertical subspace of $T_y(J^{k-1}M)$ over N. Besides, it is not hard to see that for all sections $s \in Sec(M)$ and for each $Y \in T_xN$, $Z \in Ver(J^kM)$ over N, one has

$$(T\rho^k_{k-1} - T\tilde{s}^{k-1} \circ T\alpha^k)(T\tilde{s}^k(Y)) = 0 \qquad (8)_1$$

$$(T\rho^k_{k-1} - T\tilde{s}^{k-1} \circ T\alpha^k)(Z) = T\rho^k_{k-1}(Z) \qquad (8)_2$$

DEFINITION (4). We represent (*) by θ_k and we call it k^{th} <u>order canonical form</u> on J^kM, (in the literature, we also find <u>structure</u> or <u>fundamental</u> for θ_k).

The form θ_k acts on $T(J^kM)$ with values on $Ver(J^{k-1}M)$ and so this is not a <u>real</u> but a $T(J^{k-1}M)$ - valued form. There is a somehow different but essentially the same definition for θ_k (cf. Muñoz (1984)). Let s be a section of M at a point $x \in N$ and consider the linear operator

$$(D_ks)_x : T_{\tilde{s}^{k-1}(x)}(J^{k-1}M) \to Ver_{\tilde{s}^{k-1}(x)}(J^{k-1}M)$$

defined by

$$(D_ks)_x = Id - (T\tilde{s}^{k-1} \circ T\alpha^{k-1}).$$

Then the canonical form of order k is the 1-form on J^kM defined by

$$\theta^k_{\tilde{s}^k(x)} = (D_ks)_x \circ T\rho^k_{k-1} .$$

Clearly:

$$\begin{aligned}\theta^k_{\tilde{s}^k(x)}(X) &= (D_ks)_x \circ T\rho^k_{k-1}(X) = T\rho^k_{k-1}(X) - (T\tilde{s}^{k-1} \circ T\alpha^{k-1} \circ T\rho^k_{k-1})(X) \\ &= T\rho^k_{k-1}(X) - T\tilde{s}^{k-1} \circ T\alpha^k(X)\end{aligned}$$

since on $\tilde{s}^k(x)$, $T\alpha^k = T\alpha^{k-1} \circ T\rho^k_{k-1}$. We call $D_k s$ <u>vertical differential of order</u> k of a section $s \in Sec(M)$.

Let $(x_a, y^A, z^A_{a(1)}, \ldots, z^A_{a(k)})$ be a coordinate system on J^kM induced by (x_a, y^A) on (M,p,N). If $X \in T(J^kM)$ is locally given by

$$X = X_a(\partial/\partial x_a) + X^A(\partial/\partial y^A) + \ldots + X^A_{a(k)}(\partial/\partial z^A_{a(k)})$$

then for $u \in Sec(J^kM)$

$$\theta_k(X(u(x))) =$$

$$= (X^A - z^A_a X_a) \otimes (\partial/\partial y^A) + \ldots + (X^A_{a(k-1)} - z^A_{a(k)} X_{a_k}) \otimes (\partial/\partial z^A_{a(k-1)})$$

and therefore one obtains the set of real 1-forms $\{\theta^A, \theta^A_{a(1)}, \ldots, \theta^A_{a(k-1)}\}$:

$$\left.\begin{aligned} i_X\theta^A &= X^A - z^A_a X_a & &\therefore\; \theta^A = dy^A - z^A_a dx_a \\ i_X\theta^A_a &= X^A_a - z^A_{a(2)} X_{a_2} & &\therefore\; \theta^A_a = dz^A_a - z^A_{a(2)} dx_{a_2} \\ &\vdots & & \\ i_X\theta^A_{a(k-1)} &= X^A_{a(k-1)} - z^A_{a(k)} X_{a_k} & &\therefore\; \theta^A_{a(k-1)} = dz^A_{a(k-1)} - z^A_{a(k)} dx_{a_k} \end{aligned}\right\} \quad (9)$$

REMARK (2). A sophisticated approach for a geometrical construction of the structure form can be founded in Kockynos (1979), (1982), inspired in Spencer's theory for partial differential equations.

PROPOSITION (1). <u>For every section</u> $s \in Sec(M)$, θ_k <u>is the unique form satisfying equations</u> $(8)_1$ <u>and</u> $(8)_2$.

<u>Proof</u>: Suppose that $\omega_k\colon T_z(J^kM) \to Ver_y(J^{k-1}M)$ is a form satisfying also $(8)_1$ and $(8)_2$, where $z = \tilde{s}^k(x)$, $y = \rho^k_{k-1}(z)$

for an arbitrary section $s \in Sec(M)$ and $x \in N$. If $X \in T_z(J^kM)$ then

$$X - (T\tilde{s}^k \circ T\alpha^k)(X) = Y \in Ver_y(J^{k-1}M).$$

As for vertical vector fields Z, $(\omega_k - \theta_k)(Z) = 0$, one obtains

$$(\omega_k-\theta_k)(X) = (\omega_k-\theta_k)(Y+(T\tilde{s}^k \circ T\alpha^k)(X)) =$$

$$= (\omega_k-\theta_k)(T\tilde{s}^k \circ T\alpha^k)(X) = (\omega_k-\theta_k)(T\tilde{s}^k[T\alpha^k(X)]) \overset{(8)_1}{=} 0,$$

i.e., $\omega_k = \theta_k$. □

DEFINITION (5). Let W be a submanifold of (M,p,N). We say that W is a <u>cross-section submanifold</u> of M if p/W is a bijection of W onto a submanifold of N and for every $y \in W$, $Tp: T_yW \to T_{p(y)}N$ is an isomorphism (p/W is a regular projection).

If $s \in Sec(M)$ then s defines a cross-section submanifold of M since s is a regular inclusion. We put $s(N) = N_s$ (or if $s \in Sec_V(M)$, $s(V) = V_s$). The k-jet prolongation of sections of M defines also cross-section submanifolds and in this case we put $\tilde{s}^k(N) = N_s^k$. Sometimes we identify in the following $\tilde{s}^k$ with N_s^k and we say that $\tilde{s}^k$ is a cross-section submanifold of J^kM.

The relation between k-jet prolongations and the canonical form θ_k is given by the following important

PROPOSITION (2). <u>The only cross-section submanifold of J^kM for which θ_k vanishes is the k-jet prolongation of sections of</u> $Sec(M)$.

This proposition is a consequence of the following

PROPOSITION $(2)_1$. Let $u \in \mathrm{Sec}(J^k M)$ and N_u the corresponding cross-section submanifold. Suppose that $N_u \subseteq (\beta^k)^{-1}(y)$, where $\beta^k : J^k M \to M$ is the target projection and $y \in M$. Then there is a section $s \in \mathrm{Sec}(M)$ such that $u = \tilde{s}^k$ (i.e., u is a k-jet prolongation) if and only if $\theta^A_{a(i-1)}/N_u \equiv 0$ for all $1 \le i \le k$.

Proof: If $\tilde{s}^k = u$ then from Proposition (1) $i_{T\tilde{s}^k(X)}\theta_k = 0$ and so $\theta^A_{a(i-1)}$ restricted to N_u vanishes for all $i = 1,2,\ldots,k$ (see (9)).

We show the other direction by a direct calculation. For example, let $k = 1$ and $u: N \to J^1(M)$ with $N_u \subseteq (\beta^1)^{-1}(x_a, y^A)$, (x_a, y^A) local coordinates for (M,p,N). We have

$$u(x) = (x_a, y^A, z^A_a), \qquad x = x_a \in N.$$

But $\theta^A = dy^A - z^A_a\, dx_a/N_u$ vanishes and so

$$z^A_a = \frac{dy^A}{dx_a} = d_a y^A = \frac{\partial y^A}{\partial x_a} = y^A_a$$ (see the definition of formal derivative).

Clearly $u = \tilde{s}^1$ for $s(x) = (x_a, y^A) = \beta^1 \circ u(x)$,

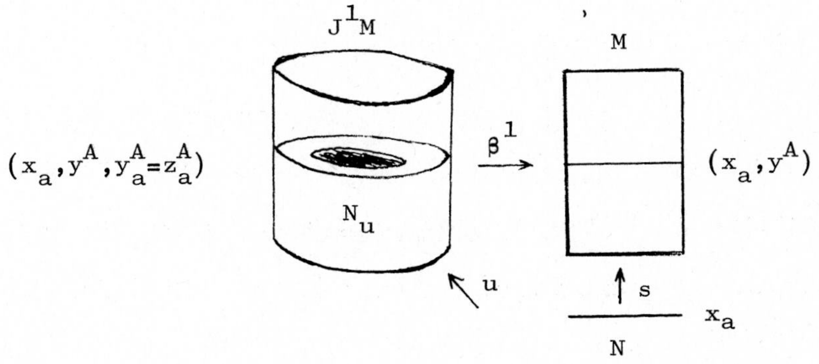

By induction on k, if we suppose that $\rho^k_{k-1}\circ u = \tilde{s}^{k-1} = j^{k-1}(\beta^{k-1}(\rho^k_{k-1}\circ u))$, $x \mapsto (x_a(y^A,\ldots,y^A_{a(k-1)})$ then $u: x \rightsquigarrow (x_a, y^A,\ldots,y^A_{a(k-1}, z^A_{a(k)})$ and since $\theta^A_{a(k-1)}$ vanishes along $N_u \subset J^kM$, one obtains

$$z^A_{a(k)} = d_{a_k} y^A_{a(k-1)} = \frac{\partial^k y^A}{\partial x_{a(k)}} = y^A_{a(k)} \quad \text{and so} \quad u = j^k(\beta^k \circ u).$$

□

REMARK (3). A proof directly involving the form θ_k may be found in Goldschmidt & Sternberg (1973). We also remark that Proposition $(2)_1$ ratifies the fact that along jet prolongations the coordinates are partial derivatives. If N_u $(= u(N)))$ is a submanifold of dimension n of J^kM given by the constraint relations

$$u^A(x_a) = y^A, \quad u^A_{a(1)}(x_a) = z^A_{a(1)},\ldots,u^A_{a(k)}(x_a) = z^A_{a(k)},$$

where $u \in \mathrm{Sec}(J^kM)$ is given by $u(x_a) = (x_a, u^A(x_a),\ldots,u^A_{a(k)}(x_a))$ then the canonical form vanishes along N_u if and only if $\partial^i u^A/\partial x_{a(i)} = u^A_{a(i)}$, $0 \le i \le k$.

3.3. <u>Higher order variational problems and exterior differential systems.</u>

Let (M,p,N) be the configuration manifold with N endowed with a volume form ω. We identify ω with $(\alpha^k)^*\omega$ since locally they have the same expression. Let $\mathfrak{L}: J^kM \to \mathbb{R}$ be a Lagrangian defined on J^kM. In the following, we will be interested in the <u>variational problem of order</u> k

$$J(s) = \int_N (\tilde{s}^k)^*\mathfrak{L}\omega = \int_{\tilde{s}^k(N)} \mathfrak{L}\omega = \int_{N^k_s} \mathfrak{L}\omega \tag{10}$$

where $s \in Sec(M)$. We take the sections in $Sec(M)$ and not in $Sec(J^kM)$ since this is an extension of the standard theory where the extremals are curves $t \mapsto (t, q(t))$ for a Lagrangian depending on $(t, q(t), dq(t)/dt)$. Recall also that in §3.1 we have taken an arbitrary fibered manifold (Q, π, N) buth here we are considering (J^kM, α^k, N) and (M, p, N). Therefore we are working with a couple (s, X) (with $s \in Sec(M)$, $X \in \mathfrak{X}(M)$) but the formalism is developed on (J^kM, α^k, N). We preserve the same assumptions about the domain of integration.

DEFINITION (6). Suppose that (A, π, B) is a fibered vector bundle. We say that a vector field $Y: A \to TA$ is <u>projectable along</u> B (or π-<u>projectable</u>) if there is a vector field $X: B \to TB$ such that for every point $q \in A$,

$$T\pi(Y(q)) = X(\pi(q)).$$

DEFINITION (7). Let $X: M \to TM$ be a vector field. The <u>k-jet prolongation</u> of X is the projectable vector field $\bar{X}^k: J^kM \to T(J^kM)$ along M (i.e. $T\beta^k(\bar{X}^k) = X$) such that $\bar{X}^k$ is an <u>infinitesimal contact transformation</u> (i.c.t.), i.e.,

$$\mathcal{L}_{\bar{X}^k} \theta^A_{a(i)} = \sum_{j \le i} f^{A/b(j)}_{B/a(i)} \theta^B_{b(j)} \qquad 0 \le i \le k-1 \tag{11}$$

(the i.c.t. assure us that the exterior differential system generated by the θ's is preserved under the action of the Lie derivative).

Formula (11) allows us to calculate explicitly $\bar{X}^k$ and the coefficients f's. For example when $k = 1$, putting $\bar{X}^1 = X + \bar{X}^A_a(\partial/\partial x_a) + X^A(\partial/\partial y^A)$, a direct calculation shows that

$$f_B^A = \partial X^A/\partial y^B - (\partial X_b/\partial y^B) z_b^A$$

and

$$\bar{X}_a^A = \frac{\partial X^A}{\partial x_a} - z_b^A \frac{\partial X_b}{\partial x_a} + z_a^B \left(\frac{\partial X^A}{\partial y^B} - \frac{\partial X_b}{\partial y^B} z_b^A\right).$$

DEFINITION (8). A section $s \in \mathrm{Sec}(M)$ is said extremal (or critical) of the variational problem (10) if for all vector fields $X: M \to TM$ we have

$$\int_N (\tilde{s}^k)^* \mathcal{L}_{\bar{X}^k} \mathfrak{L}\omega = 0 \tag{12}$$

(or

$$\int_{N_s^k} \mathcal{L}_{\bar{X}^k} \mathfrak{L}\omega = 0).$$

REMARK (4). In Muñoz (1984) the reader can find a proof that for every vector field X on M there is a unique i.c.t. $\bar{X}^k$ projectable onto X. In fact Muñoz's approach extends Definition (7) to the vector valued structure form θ_k by imposing the condition $\mathcal{L}_{\bar{X}^k} \theta_k = F\,\theta_k$, where F is an endomorphism of the vertical vector bundle $\mathrm{Ver}(J^{k-1}M)$.

Let us consider again $k = 1$,

$$\theta^A = dy^A - z_b^A dx_b, \quad \omega_c = (-1)^{c-1} dx_1 \wedge \ldots \wedge \widehat{dx_c} \wedge \ldots \wedge dx_n .$$

Since only the vertical part of the vector field X on M is relevant for the calculations we restrict our attention to $\mathrm{Ver}\, M$. The 1-jet prolongation of a vertical vector field X on M is

$$\bar{X}^1 = X^A (\partial/\partial y^A) + \left(\frac{\partial X^A}{\partial x_b} + \frac{\partial X^A}{\partial y^B} y_b^B\right)(\partial/\partial z_b^A).$$

Now,

$$\mathcal{L}_{\bar{X}^1}(\Omega\omega) = \mathcal{L}_{\bar{X}^1}(\Omega)\omega = i_{\bar{X}^1}\, d\Omega \wedge \omega$$

$$= X^A \frac{\partial\Omega}{\partial y^A}\omega + \left(\frac{\partial X^A}{\partial x_b} + \frac{\partial X^A}{\partial y^B} z^B_b\right)\frac{\partial\Omega}{\partial z^A_c}\, dx_b \wedge \omega_c\,, \qquad (dx_b \wedge \omega_c = \delta^b_c\omega).$$

Taking into account that

$$\frac{\partial\Omega}{\partial z^A_c}\, d(X^A) \wedge \omega_c = d\left(\frac{\partial\Omega}{\partial z^A_c} X^A\omega_c\right) - X^A\, d\left(\frac{\partial\Omega}{\partial z^A_c}\right) \wedge \omega_c$$

and

$$\frac{\partial X^A}{\partial y^B}\frac{\partial\Omega}{\partial z^A_c} y^B_b\, dx_b \wedge \omega_c = \frac{\partial X^A}{\partial y^B}\frac{\partial\Omega}{\partial z^A_c}(dy^B \wedge \omega_c - \theta^B \wedge \omega_c),$$

we see that

$$\mathcal{L}_{\bar{X}^1}(\Omega\omega) = X^A \frac{\partial\Omega}{\partial y^A}\omega + \left(\frac{\partial X^A}{\partial x_b}\frac{\partial\Omega}{\partial z^A_c}\right)dx_b \wedge \omega_c + \frac{\partial X^A}{\partial y^B}\frac{\partial\Omega}{\partial z^A_c}\, dy^B \wedge \omega_c -$$

$$- \frac{\partial X^A}{\partial y^B}\frac{\partial\Omega}{\partial z^A_c}\theta^B \wedge \omega_c = X^A\frac{\partial\Omega}{\partial y^A}\omega + \frac{\partial\Omega}{\partial z^A_c}\left(\frac{\partial X^A}{\partial x_b}dx_b + \frac{\partial X^A}{\partial y^B}dy^B\right) \wedge \omega_c -$$

$$- \frac{\partial X^A}{\partial y^B}\frac{\partial\Omega}{\partial z^A_c}\theta^B \wedge \omega_c = X^A\frac{\partial\Omega}{\partial y^B}\omega + \frac{\partial\Omega}{\partial z^A_c}dX^A \wedge \omega_c - \frac{\partial X^A}{\partial y^B}\frac{\partial\Omega}{\partial z^A_c}\theta^B \wedge \omega_c$$

$$= X^A\frac{\partial\Omega}{\partial y^B}\omega + d\left(\frac{\partial\Omega}{\partial z^A_c}X^A\omega_c\right) - X^A d\left(\frac{\partial\Omega}{\partial z^A_c}\right) \wedge \omega_c - \frac{\partial X^A}{\partial y^B}\frac{\partial\Omega}{\partial z^A_c}\theta^B \wedge \omega_c\,.$$

Therefore

$$\mathcal{L}_{\bar{X}^1}(\Omega\omega) = X^A\left(\frac{\partial\Omega}{\partial y^B}\omega - d\left(\frac{\partial\Omega}{\partial z^A_c}\right) \wedge \omega_c\right) + d\left(\frac{\partial\Omega}{\partial z^A_c}X^A\omega_c\right) - \frac{\partial X^A}{\partial y^B}\frac{\partial\Omega}{\partial z^A_c}\theta^B \wedge \omega_c.$$

If we consider the problem (12) for (vertical) vector fields for which $\bar{X}^1/\partial N^1_s \equiv 0$ then the use of Stokes' theorem for (12) gives along 1-jet prolongations $\tilde{s}^1$ (recall that in such a case $\theta^B \equiv 0$),

$$X^A\left(\frac{\partial \mathfrak{L}}{\partial y^B}\,\omega - d\left(\frac{\partial \mathfrak{L}}{\partial y^A_c}\right) \wedge \omega_c\right) = 0.$$

But from the definition of total derivative along 1-jet prolongations we have $dF \wedge \omega_c = d_c F\, dx_c \wedge \omega_c = d_c F\omega$ and hence

$$X^A\left[\frac{\partial \mathfrak{L}}{\partial y^B} - d_c\left(\frac{\partial \mathfrak{L}}{\partial y^A_c}\right)\right]\omega = 0 \Rightarrow \frac{\partial \mathfrak{L}}{\partial y^B} - \frac{d}{dx_c}\left(\frac{\partial \mathfrak{L}}{\partial y^A_c}\right) = 0,$$

that is, the Euler-Lagrange field equations.

With enough patience we may reproduce the above calculation for any finite $k > 1$ and therefore, under the above assumptions on X and $\bar{X}^k$, we see that the extremals of (12) are those sections $s \in \mathrm{Sec}(M)$ for which $(\delta\mathfrak{L}/\delta y^A)(\tilde{s}^{2k}(x_a)) = 0$, where the symbol $\delta/\delta y^A$ is defined by $(5)_1$ in §2.2.

We have pointed out in the introduction of the present chapter that in the ordinary situation variational problems of order 1 may be replaced by variational problems of order zero with the help of the Poincaré-Cartan form. We have remarked also that the Cartan approach must be studied carefully since there is no canonical extension of the standard procedure to higher order variational problems in many independent variables.

Let us show some of the problems which arise in higher order theories. For example, take $k = 2$ and consider the following forms

$$\left.\begin{aligned} G_2 &= g^a_A\,\theta^A \wedge \omega_a + g^{ab}_A\,\theta^A_a \wedge \omega_b \\ \rho &= U^a\,dx_a + U_A\,dy^A + U^a_A\,dz^A_a + U^{ab}_A\,dz^A_{ab} \\ v_2 &= dG_2 + \rho \wedge \omega \end{aligned}\right\} \qquad (13)$$

where θ^A, θ^A_a are the real components of θ_2, $\omega_a = (-1)^{(a-1)} dx_1 \wedge \ldots \wedge \widehat{dx_a} \wedge \ldots \wedge dx_n$, ω is the volume form injected into J^2M and the functions $g^a_A, g^{ab}_A, \ldots, U^{ab}_A$ <u>are supposed defined</u> on J^2M. We suppose also that $1 \leq a \leq b \leq n$.

PROPOSITION (3). <u>Let</u> Y <u>be a vertical vector field on</u> J^2M <u>over</u> N. <u>If</u>

$$i_Y v_2 = 0 \tag{14}$$

<u>at a point</u> $p \in J^2M$, <u>then</u>

$$\left.\begin{aligned} U^a_A - g^a_A - d_b(g^{ab}_A) &= 0 \\ U_A - d_a(g^a_A) &= 0 \\ U^{ab}_A - g^{ab}_A &= 0 \end{aligned}\right\} \qquad (15)_1$$

$$\begin{pmatrix} \partial g^a_A/\partial y^B & \partial g^{ab}_A/\partial y^B \\ \partial g^a_A/\partial z^B_c & \partial g^{ab}_A/\partial z^B_c \\ \partial g^a_A/\partial z^B_{cd} & \partial g^{ab}_A/\partial z^B_{cd} \end{pmatrix} \begin{pmatrix} \dfrac{\partial y^A}{\partial x_a} - z^A_a \\ \\ \dfrac{\partial z^A_b}{\partial x_a} - z^A_{ab} \end{pmatrix} \qquad (15)_2$$

The proof is obtained from a straightforward calculation. The development of (14) gives a certain number of expressions which can be rearranged as coefficients of the components functions of the vector field Y. We will see this below for a somewhat different situation (see Proposition $(3)_1$).

From equations $(15)_2$ one obtains

$$d_a(g^a_A) - d_a(U^a_A) + d_a d_b(g^{ab}_A) = U_A - d_a(U^a_A) + d_a d_b(U^{ab}_A) = 0, \tag{16}$$

where the formal derivative of the first equality in $(15)_2$ was taken. If we suppose now that $\rho = d\Omega$, for a function

$\mathfrak{L}: J^2 M \rightarrow \mathbb{R}$,   then (16) assumes the form

$$\frac{\partial \mathfrak{L}}{\partial y^A} - \frac{d}{dx_a}\left(\frac{\partial \mathfrak{L}}{\partial z_a^A}\right) + \frac{d^2}{dx_a dx_b}\left(\frac{\partial \mathfrak{L}}{\partial z_{ab}^A}\right) = 0 \tag{17}$$

which has a formal resemblance with the Euler-Lagrange equations. In fact if $u \in \mathrm{Sec}(J^2 M)$ is such that $u = \tilde{s}^2$ for some $s \in \mathrm{Sec}(M)$ then $p = \tilde{s}^2(x_a)$ and along N_s^2 equations (17) are Euler-Lagrange since $z_a^A = \dfrac{\partial u^A}{\partial x_a}$, $\partial z_{ab}^A = \dfrac{\partial^2 u^A}{\partial x_a \partial x_b}$, where $u^A(x_a) = y^A$, $u_b^A(x_a) = z_b^A$. This last condition on jet prolongation may be obtained directly from the second system of equations $(15)_2$. Since the functions g_A^a, g_A^{ab} must depends on the variables y^B, if we assume that each sub-matrix in the left hand side matrix is of maximal rank, then the 2-jet prolongation condition is fulfilled. In other words, this condition is obtained when the system has unique trivial solution. This is a sufficient condition for jet prolongations since we may have degeneracy along N_s^2 for the arbitrary functions g_A^a, g_A^{ab}.

COROLLARY. With the conditions of Proposition (3), now with $\rho = d\mathfrak{L}$, where $\mathfrak{L}: J^2 M \rightarrow \mathbb{R}$ is a smooth function, equation (14) generates the Euler-Lagrange field equations when the system $(15)_2$ has unique trivial solution.

If we examine $(15)_2$, then we may define

$$g_A^{ab} = \partial \mathfrak{L}/\partial y_{ab}^A ; \quad g_A^a = \partial \mathfrak{L}/\partial y_a^A - d_a(\partial \mathfrak{L}/\partial y_{ab}^A) \tag{18}$$

when $(15)_2$ has unique trivial solution (and therefore the 2-jet prolongation condition is fulfilled). Putting

$$\mathcal{H} = g_A^a y^A + g_A^{ab} y_{ab}^A - \mathfrak{L}(x_a, y^A, y_a^A, y_{ab}^A),$$

we get

$$\left.\begin{aligned} &\frac{\partial \mathcal{H}}{\partial y^A} = \frac{\partial \mathfrak{L}}{\partial y^A} = -d_a(g^a_A); \qquad \frac{\partial \mathcal{H}}{\partial y^A_a} = g^a_A - \frac{\partial \mathfrak{L}}{\partial y^A_a} = -d_b\left(\frac{\partial \mathfrak{L}}{\partial y^A_{ab}}\right) \\ &\frac{\partial \mathcal{H}}{\partial g^a_A} = y^A_a ; \qquad\qquad \frac{\partial \mathcal{H}}{\partial g^{ab}_A} = y^A_{ab} \end{aligned}\right\} \quad (19)_1$$

with

$$\partial \mathcal{H}/\partial y^A_{ab} = g^{ab}_A - \partial \mathfrak{L}/\partial y^A_{ab} = 0 \qquad (19)_2$$

The reader may think that $(19)_1$ are the Hamilton equations, with (18) being the Jacobi-Ostrogradsky momenta. However this is only a formal calculation since (18) contradicts the fact that v_2 is a form on J^2M. From (18) we see that

$$g^{ab}_A = g^{ab}_A\ (x_a, y^A, y^A_a, y^A_{ab})$$

and so g^{ab}_A is defined on J^2M. But

$$g^a_A = g^a_A\ (x_a, y^A, y^A_a, y^A_{ab}, y^A_{abc})$$

and g^a_A must be defined on J^3M. However, we may express Proposition (3) in a different form. Let us recall that the <u>partial derivative</u> of the form $v = \Sigma a_{i(s)}\, \sigma_{i_1} \wedge \ldots \wedge \sigma_{i_s}$ with respect to σ_{i_r} is the form

$$\frac{\partial v}{\partial \sigma_{i_r}} = (-1)^{r-1}\, \sigma_{i_1} \wedge \ldots \wedge \widehat{\sigma_{i_r}} \wedge \ldots \wedge \sigma_{is}; \quad \frac{\partial v}{\partial \sigma_j} = 0 \text{ if } j \neq i_r,\ 1 \leq r \leq s.$$

Let $\mathfrak{L}: J^2M \to \mathbb{R}$ be a Lagrangian and consider the following set of forms:

$$\left.\begin{aligned}
&\theta^A = dy^A - z^A_a dx_a;\quad \theta^A\wedge\omega = dy^A\wedge\omega;\quad \theta^A\wedge\omega_a = dy^A\wedge z^A_a\omega\\
&\theta^A_a = dz^A_a - z^A_{ab}dx_b;\quad \theta^A_a\wedge\omega = dz^A_a\wedge\omega;\quad \theta^A_a\wedge\omega_b = dz^A_a\wedge\omega_b - z^A_{ab}\omega\\
&\theta^A_{ab} = dz^A_{ab} - z^A_{abc}dx_c;\quad \theta^A_{ab}\wedge\omega = dz^A_{ab}\wedge\omega;\quad \theta^A_{ab}\wedge\omega_c = dz^A_{ab}\wedge\omega_c - z^A_{abc}\omega\\
&\theta^A_{abc} = dz^A_{abc} - z^A_{abcd}dx_d;\quad \theta^A_{abc}\wedge\omega = dz^A_{abc}\wedge\omega;\quad \theta^A_{abc}\wedge\omega_d = dz^A_{abc}\wedge\omega_d - z^A_{abcd}\omega
\end{aligned}\right\}(20)_1$$

$$\left.\begin{aligned}
&d\theta^A = -dz^A_a\wedge dx_a;\quad d\theta^A\wedge\omega_a = -\theta^A_a\wedge\omega\\
&d\theta^A_a = -dz^A_{ab}\wedge dx_b;\quad d\theta^A_a\wedge\omega_b = -\theta^A_{ab}\wedge\omega
\end{aligned}\right\}(20)_2$$

Suppose that

$$\Omega_2 = g^a_A\,\theta^A\wedge\omega_a + g^{ab}_A\,\theta^A_a\wedge\omega_b + \mathcal{L}\omega$$

is a form defined on J^3M, where now $g^a_A : J^3M \to \mathbb{R}$, $g^{ab}_A : J^2M \to \mathbb{R}$ are smooth functions. If we take $d\Omega_2$ $(=v_2)$ and if we apply relations $(20)_1$ and $(20)_2$ in $d\Omega_2$, a direct calculation shows that

$$d\Omega_2 = \theta^A\wedge\left(\frac{\partial\mathcal{L}}{\partial y^A}\omega - dg^a_A\wedge\omega_a\right) + \theta^A_a\wedge$$
$$\wedge\left[\left(\frac{\partial\mathcal{L}}{\partial z^A_a} - g^a_A\right)\omega - dg^{ab}_A\wedge\omega_b\right] + \theta^A_{ab}\wedge\left(\frac{\partial\mathcal{L}}{\partial z^A_{ab}} - g^{ab}_A\right)\omega .$$

Now, as g^a_A (resp. g^{ab}_A) is defined on J^3M (resp. J^2M), we have, with the help of $(20)_1$

$$dg^a_A\wedge\omega_a = \frac{\partial g^a_A}{\partial x_a}\omega + \frac{\partial g^a_A}{\partial y^B}dy^B\wedge\omega_a + \frac{\partial g^a_A}{\partial z^B_c}dz^B_c\wedge\omega_a +$$
$$+ \frac{\partial g^a_A}{\partial z^B_{cd}}dz^B_{cd}\wedge\omega_a + \frac{\partial g^a_A}{\partial z^B_{cde}}dz^B_{cde}\wedge\omega_a =$$
$$= \frac{\partial g^a_A}{\partial x_a}\omega + \frac{\partial g^a_A}{\partial y^B}(\theta^B\wedge\omega_a + z^B_a\omega) + \frac{\partial g^a_A}{\partial z^B_b}(\theta^B_c\wedge\omega_a + z^B_{ac}\omega) +$$

$$+ \frac{\partial g^a_A}{\partial z^B_{cd}} (\theta^B_{cd} \wedge \omega_a + z^B_{acd}\omega) + \frac{\partial g^a_A}{\partial z^B_{cde}} (\theta^B_{cde} \wedge \omega_a + z^B_{acde}\omega),$$

therefore,

$$dg^a_A \wedge \omega_a = (d_a g^a_A)\omega + \frac{\partial g^a_A}{\partial y^B} \theta^B \wedge \omega_a + \frac{\partial g^a_A}{\partial z^B_c} \theta^B_c \wedge \omega_a + \frac{\partial g^a_A}{\partial z^B_{cd}} \theta^B_{cd} \wedge \omega_a + $$
$$+ \frac{\partial g^a_A}{\partial z^B_{cde}} \theta^B_{cde} \wedge \omega_a$$

$$dg^{ab}_A \wedge \omega_b = (d_b g^{ab}_A)\omega + \frac{\partial g^{ab}_A}{\partial y^B} \theta^B \wedge \omega_b + \frac{\partial g^{ab}_A}{\partial z^B_c} \theta^B_c \wedge \omega_b + \frac{\partial g^{ab}_A}{\partial z^B_{cd}} \theta^B_{cd} \wedge \omega_b$$

and

$$d\Omega_2 = \theta^A \wedge (\frac{\partial \Omega}{\partial y^A} - d_a g^a_A)\omega -$$

$$- \theta^A \wedge (\frac{\partial g^a_A}{\partial y^B} \theta^B \wedge \omega_a + \frac{\partial g^a_A}{\partial z^B_c} \theta^B_c \wedge \omega_a + \frac{\partial g^a_A}{\partial z^B_{cd}} \theta^B_{cd} \wedge \omega_a + \frac{\partial g^a_A}{\partial z^B_{cde}} \theta^B_{cde} \wedge \omega_a)$$

$$+ \theta^A_a \wedge \left[(\frac{\partial \Omega}{\partial z^A_a} - g^a_A - d_b g^{ab}_A)\omega\right] -$$

$$- \theta^A_a \wedge (\frac{\partial g^{ab}_A}{\partial y^B} \theta^B \wedge \omega_b + \frac{\partial g^{ab}_A}{\partial z^B_c} \theta^B_c \wedge \omega_b + \frac{\partial g^{ab}_A}{\partial z^B_{cd}} \theta^B_{cd} \wedge \omega_b) + \theta^A_{ab} \wedge (\frac{\partial \Omega}{\partial z^A_{ab}} - g^{ab}_A)\omega$$

$$= \theta^A \wedge (\frac{\partial \Omega}{\partial y^A} - d_a g^a_A)\omega - \{-\theta^B (\frac{\partial g^a_A}{\partial y^B} \theta^A \wedge \omega_a + \frac{\partial g^{ab}_A}{\partial y^B} \theta^A_a \wedge \omega_b) -$$

$$- \theta^B_c (\frac{\partial g^a_A}{\partial z^B_c} \theta^A \wedge \omega_a + \frac{\partial g^{ab}_A}{\partial z^B_c} \theta^A_a \wedge \omega_b) - \theta^B_{cd} (\frac{\partial g^a_A}{\partial z^B_{cd}} \theta^A \wedge \omega_a + \frac{\partial g^{ab}_A}{\partial z^B_{cd}} \theta^A_a \wedge \omega_b)\}$$

$$+ \theta^B_{cde} \wedge (\frac{\partial g^a_A}{\partial z^B_{cde}}) \theta^A \wedge \omega_a + \theta^A_a \left[(\frac{\partial \Omega}{\partial z^A_a} - g^a_A - d_b g^{ab}_A)\omega\right] + \theta^A_{ab} (\frac{\partial \Omega}{\partial z^A_{ab}} - g^{ab}_A)\omega.$$

If we suppose that

$$\partial d\Omega_2 / \partial \theta^A = \ldots = \partial d\Omega_2 / \partial \theta^B = \ldots = \partial d\Omega_2 / \partial \theta^A_{ab} = 0 \tag{21}$$

then

$$\left.\begin{aligned}
&\frac{\partial \mathfrak{L}}{\partial z^A_a} - g^a_A - d_b g^{ab}_A = 0\\
&\frac{\partial \mathfrak{L}}{\partial y^A} - d_a g^a_A = 0\\
&\frac{\partial \mathfrak{L}}{\partial z^A_{ab}} - g^{ab}_A = 0
\end{aligned}\right\}(22)_1
\qquad
\left.\begin{aligned}
&\frac{\partial g^a_A}{\partial y^B}\,\theta^A\wedge\omega_a + \frac{\partial g^{ab}_A}{\partial y^B}\,\theta^A_a\wedge\omega_b = 0\\
&\frac{\partial g^a_A}{\partial z^B_c}\,\theta^A\wedge\omega_a + \frac{\partial g^{ab}_A}{\partial z^B_c}\,\theta^A_a\wedge\omega_b = 0\\
&\frac{\partial g^a_A}{\partial z^B_{cd}}\,\theta^A\wedge\omega_a + \frac{\partial g^{ab}_A}{\partial z^B_{cd}}\,\theta^A_a\wedge\omega_b = 0\\
&\frac{\partial g^a_A}{\partial z^B_{cde}}\,\theta^A\wedge\omega_a = 0
\end{aligned}\right\}(22)_2$$

Let $u\colon N \to J^3M$ be a section; $u(x) = (x_a, u^A(x), u^A_a(x), u^A_{ab}(x), u^A_{abc}(x)) \in u(N)$, $x \in N$. Then along $u(N)$ we have

$$\theta^A = \left(\frac{\partial u^A}{\partial x_a} - z^A_a\right)dx_a\,; \qquad \theta^A_a = \left(\frac{\partial u^A_a}{\partial x_b} - z^A_{ab}\right)dx_b$$

and $(22)_2$ takes the form

$$\begin{pmatrix}
\dfrac{\partial g^a_A}{\partial y^B} & \dfrac{\partial g^{ab}_A}{\partial y^B}\\[2ex]
\dfrac{\partial g^a_A}{\partial z^B_c} & \dfrac{\partial g^{ab}_A}{\partial z^B_c}\\[2ex]
\dfrac{\partial g^a_A}{\partial z^B_{cd}} & \dfrac{\partial g^{ab}_A}{\partial z^B_{cd}}\\[2ex]
\dfrac{\partial g^a_A}{\partial z^B_{cde}} & \bigcirc
\end{pmatrix}
\begin{pmatrix}
\dfrac{\partial u^A}{\partial x_a} - z^A_a\\[2ex]
\dfrac{\partial u^A_a}{\partial x_b} - z^A_{ab}
\end{pmatrix} = 0 \qquad (22)_3$$

PROPOSITION $(3)_1$. *Suppose that* (21) *is verified and let* $u\colon N \to J^3M$ *be a section. Then along* $u(N)$ *one has the systems* $(22)_1$ *and* $(22)_3$.

The extension for higher order Lagrangians can be similarly obtained for a special Ω_k defined locally on $J^{2k-1}M$ by

$$\Omega_k = g_A^{a_1}\, \theta^A \wedge \omega_{a_1} + \dots + g_A^{a_1 a_2 \dots a_k}\, \theta^A_{a_1 \dots a_{k-1}} \wedge \omega_k + \mathfrak{L}\omega \qquad (23)$$

where the functions g's are defined respectively on $J^{2k-1}M$, $J^{2k-2}M$, etc. Also, these functions are such that $a_1 \leq a_2 \leq \dots \leq a_k$ and so we must be careful when we develop (23) in the summation. The extension of Proposition $(3)_1$ gives a generalized systems of equations of type

$$\left.\begin{aligned}
&\frac{\partial \mathfrak{L}}{\partial z^A_{a_1 \dots a_{k-1}}} - g_A^{a_1 \dots a_{k-1}} - d_{a_k}\left(g_A^{a_1 \dots a_k}\right) = 0 \\
&\qquad \vdots \qquad\qquad \vdots \qquad\qquad \vdots \\
&\frac{\partial \mathfrak{L}}{\partial z^A_{a_1}} - g_A^{a_1} - d_{a_2}\left(g_A^{a_1 a_2}\right) = 0 \\
&\frac{\partial \mathfrak{L}}{\partial y^A} - d_{a_1}\left(g_A^{a_1}\right) = 0 \\
&\frac{\partial \mathfrak{L}}{\partial z^A_{a_1 \dots a_k}} - g_A^{a_1 \dots a_k} = 0
\end{aligned}\right\} \quad (23)_1$$

$$\begin{pmatrix}
\dfrac{\partial g_A^{a_1}}{\partial y^B} & \dfrac{\partial g_A^{a_1 a_2}}{\partial y^B} & \cdots & \dfrac{\partial g^{a_1 \cdots a_k}}{\partial y^B} \\
\vdots & \vdots & & \vdots \\
\dfrac{\partial g_A^{a_1}}{\partial z^B_{a_1 \cdots a_k}} & \dfrac{\partial g_A^{a_1 a_2}}{\partial z^B_{a_1 \cdots a_k}} & \cdots & \dfrac{\partial g^{a_1 \cdots a_k}}{\partial z^B_{a_1 \cdots a_k}} \\
\dfrac{\partial g_A^{a_1}}{\partial z^B_{a_1 \cdots a_k}} & \dfrac{\partial g_A^{a_1 a_2}}{\partial z^B_{a_1 \cdots a_{k+1}}} & \cdots & O \\
\vdots & \vdots & & \vdots \\
\dfrac{\partial g_A^{a_1}}{\partial z^B_{a_1 \cdots a_{2k-1}}} & O & & O
\end{pmatrix}
\begin{pmatrix}
\dfrac{\partial y^A}{\partial x_{a_1}} - z^A_{a_1} \\
\\
\dfrac{\partial z^A_{a_1 \cdots a_{k-1}}}{\partial x_{a_k}} - a^A_{a_1 \cdots a_k}
\end{pmatrix} = 0$$

$(23)_2$

We call this matrix by <u>characteristic matrix</u>. When $(23)_2$ has only trivial solution, a section $u: N \to J^{2k-1}M$ will be k-jet prolongation of a section $s: N \to M$ (sometimes we say that u is k-holonomic). Also, it is possible to show that from $(23)_1$ one obtains the generalized Euler-Lagrange field equations (5) of §2.2 along the k-jet of $S: N \to M$. Unfortunately this is insufficient to obtain the corresponding Hamiltonian theory.

Since the early work of Dedecker (1950), (1953) many investigations has been developed with the intention of extending the variational formalism of Cartan (1922) to theories involving higher order derivatives with many independent variables, i.e., of establishing a similar relation between a kth order variational problem given by a (higher order) Poincaré-Cartan form. We may say that the main question in such study is precisely to know <u>if such zeroth order problem</u>

exists in a way that both variational problems are equivalent.

If $\dim N = 1$ and $k > 1$ is finite or if $\dim N > 1$ is finite and $k = 1$ then the answer is yes; there is a unique Poincaré-Cartan form and it is possible to establish the equivalence between both variational formalisms, (see Dedecker (1950), (1977), Garcia (1968), Goldschmidt & Sternberg (1973), Kolár (1983), Krupka (1973)). More precisely, if $k = 1$, $\dim N > 1$, for example, then "$u \in \mathrm{Sec}(J^1M)$ is an extremal of the integral

$$\int_N u^*\Omega_1$$

(where Ω_1 is the Poincaré-Cartan form globally constructed on J^1M) if and only if u is of the form $\tilde{s}^1$, where s is an extremal of (10)". Locally the form Ω_1 is defined by

$$\Omega_1 = g^a_A \, \theta^A \wedge \omega_a + \mathfrak{L}\omega$$

where $\mathfrak{L}: J^1M \to \mathbb{R}$ is the Lagrangian. When $\mathfrak{L}$ is regular, $g^a_A = \partial\mathfrak{L}/\partial y_a$.

In a situation involving many independent variables with higher order derivatives, the Poincaré-Cartan form is not uniquely determined and admits many local expressions (recall that the zeroth order problem will be of type $\int_N u^*\Omega_k$, where Ω_k is the (higher order) Poincaré-Cartan form, to be defined). For example, if we take $k = 2$, then $d\Omega_2 = dG_2 + d\mathfrak{L} \wedge \omega$ and we may have $G_2 = g^a_A\theta^A \wedge \omega_b + g^{ab}_A \theta^A_a \wedge \omega_b$, $G_2 = g^a_A \, \theta^A \wedge \omega_a + g^{ab}_A \, \theta^A_b \wedge \omega_a$. If we eliminate the restriction $a \le b$, then equation $(22)_1$ is modified; the last equality in $(22)_1$ reads $\partial\mathfrak{L}/\partial z^A_{aa} = g^{aa}_A$, $\partial\mathfrak{L}/\partial z^A_{ab} - g^{ab}_A - g^{ba}_A = 0$,

$a \neq b$.

Another problem concerning the Hamiltonization of higher order Lagrangians is the question of the conditions for a section u to be an extremal such that u is the jet prolongation of a section which is an extremal of the kth order variational problem $\Omega\omega$. As we have seen, for our particular situation $(a_1 \leq a_2 \leq \ldots \leq a_k)$ even when the system $(23)_2$ admits the trivial solution an extremal $u \in Sec(J^{2k-1}u)$ is the k-jet prolongation of some section $s \in Sec(M)$ but not necessarily $(2k-1)$ jet prolongation of such s, and this condition is essential. For the general situation the problem is clearly more complicated since many relations between the coefficients of the form G_k must be verified. Finally, it remains the problem of defining a "higher order Legendre transformation".

Let us see an example due to Dedecker (1984)(a), (1984)(b), showing that in general it is untrue the following assertion: "a section is an extremal of a variational problem of order zero if and only if it is the $(2k-1)$ jet prolongation of a section which is an extremal of an kth order variational problem", (see Francaviglia & Krupka (1982)).

Suppose that $k = \dim N = 2$. Then it is possible to show that the sections $s \in Sec(M)$ which are extremals of the 2nd order problem

$$I_2 = \int_{N_s} \Omega\omega$$

have the property that $\tilde{s}^3$ is extremal of the $zero^{th}$ order problem

$$I_o = \int_{N_s^3} G_2 + \mathfrak{L}\omega$$

if and only if the relations

$$\left.\begin{aligned} g_A^{11} = (\partial\mathfrak{L}/\partial z_{11}^A);\quad g_A^{22} = (\partial\mathfrak{L}/\partial z_{22}^A);\quad g_A^{12}+g_A^{21} = (\partial\mathfrak{L}/\partial z_{12}^A) \\ g_A^{a} = (\partial\mathfrak{L}/\partial y_a^A) - d_b(g_A^{ab}) \end{aligned}\right\}(24)$$

are verified; $g_A^{ab}: J^3M \to \mathbb{R}$, $g_A^a: J^2M \to \mathbb{R}$ are functions such that $G_2 = g_A^a\,\theta^A \wedge \omega_a + g_A^{ab}\,\theta_a^A \wedge \omega_b$. The general solution of (24) is given by

$$g_A^{12} = (1/2)(\partial\mathfrak{L}/\partial z_{12}^A) + F_A;\quad g_A^{21} = (1/2)(\partial\mathfrak{L}/\partial z_{12}^A) - F_A,$$

where $F_A: J^2M \to \mathbb{R}$ is a smooth function. Now, let us consider the Lagrangian

$$\mathfrak{L} = (1/2)[(z_{11}^A)^2 + (z_{12}^A)^2 + (z_{22}^A)^2].$$

Then when $F_A = 0$ (or $F_A = \pm(1/2)(\partial\mathfrak{L}/\partial z_{12}^A)$) one has that

i) If $u \in \mathrm{Sec}(J^2M)$ is an extremal of

$$I_o = \int_N u^*(G_2+\mathfrak{L}\omega)$$

then $u = \tilde{s}^2 = j^2(\beta^2 \circ u)$ and s is an extremal of I_2;

ii) If $s \in \mathrm{Sec}(M)$ is an extremal of I_2 then s admits an infinite number of prolongations u, extremals of I_o, which are 2-jet prolongations (but not 3-jet prolongations) depending on two arbitrary functions of two variables.

REMARK (5). (i) In Dedecker's example - according our definition - the Lagrangian is "degenerated". In fact if we choose $F_A = 0$ then $g_A^{11} = z_{11}^A$, $g_A^{22} = z_{22}^A$, $g_A^{12} = z_{12}^A$ and

$g_A^{21} = z_{12}^A = g_A^{12}$. Therefore

$$\left(\frac{\partial^2 \mathfrak{L}}{\partial z_{ab}^A \partial z_{cd}^B}\right) = \begin{pmatrix} 1 & 0 \\ 0 & 1 \end{pmatrix},$$

but

$$g_A^1 = -\frac{d}{dx_1}(g_A^{11}) - \frac{d}{dx_2}(g_A^{12}) = 0$$

$$g_A^2 = -\frac{d}{dx_2}(g_A^{21}) - \frac{d}{dx_2}(g_A^{22}) = 0$$

(ii) Garcia and Muñoz (1983) proposed an intrinsical version for $k > 1$ and $\dim N > 1$ (see also Muñoz (1983), (1984)). A constructive method is proposed for the Poincaré-Cartan form. However such a form depends on a couple of linear connections defined on the base manifold N and on the vertical vector bundle of M. It is showed that if λ and λ' are two Poincaré-Cartan forms then they differ by a term which is related with the canonical form of order $(k-1)$, θ_{k-1}. We also indicate to the reader the article by Horak & Kolár (1983) where a different point of view is adopted.

3.4. Two attempts for the Hamiltonization of the field equations.

As it is well known in Classical Mechanics involving only first order derivatives, regular Lagrangians systems can be expressed into Hamiltonian form. If L is a regular Lagrangian defined on the tangent bundle TW of some configuration manifold W then its Hamiltonian counterpart is defined on the dual manifold T^*W and the Legendre transformation $Leg: TW \to T^*W$ "replaces" the variables (q_i, v_i) of TW

by the variables (q_i, p_i) of T^*W, where $p_i = \partial\mathfrak{L}/\partial v_i$. Of course, the same procedure may be adapted for regular Lagrangians in Field Theory. However when we are in the presence of a theory which involves higher order derivatives there is no canonical way of putting the Lagrangian equations into Hamiltonian form, in the above sense.

If $k = 2$, for example, then we may put the Euler-Lagrange equations in the form

$$\frac{\partial\mathfrak{L}}{\partial y^A} - \frac{d}{dx_a}\left[\frac{\partial\mathfrak{L}}{\partial y^A_a} - \frac{d}{dx_b}\left(\frac{\partial\mathfrak{L}}{\partial y^A_{ab}}\right)\right] = 0$$

and this suggests, as we have seen before, the definition

$$p_{A/a} = \frac{\partial\mathfrak{L}}{\partial y^A_a} - \frac{d}{dx_b}\left(\frac{\partial\mathfrak{L}}{\partial y^A_{ab}}\right); \qquad p_{A/ab} = \frac{\partial\mathfrak{L}}{\partial y^A_{ab}}$$

for the momenta. But it remains a question: what is the "good" definition for the Legendre transformation (or if we wish, what are the regularity conditions for $\mathfrak{L}: J^2M \to \mathbb{R}$) ? The bundle J^kM, $k > 1$ has no dual bundle and therefore it is impossible to extend the usual procedure for this situation. It also remains, as a consequence of this problem, to specify whom is "canonically" conjugate to whom?

We will be interested in the following two possibilities:

H(1) - For a Lagrangian density of order k, $\mathfrak{L}^k: J^kM \to \mathbb{R}$, if we consider the 1-jet prolongation $J^1(J^{k-1}M)$ then it is possible to develop the Hamiltonian theory over the dual manifold $J^{1*}(J^{k-1}M)$;

H(2) - For a Lagrangian density defined on J^kM, we may try to develop a Hamiltonian theory in the jet manifold $J^{2k-1}M$.

The first point of view has the clear intention of giving, as far as possible, a "natural" extension of the standard procedure to higher order systems. However this is an indirect procedure since such Hamiltonization on $J^{1*}(J^{k-1}M)$ needs to be developed in correspondence with a Lagrangian formalism on $J^1(J^{k-1}M)$ and not on J^kM. In fact we are considering a theory of order one involving a large number of, as we say in physics, "spatial variables".

The second alternative is the direct procedure mentioned before and, apart from the problems underlying the geometrical variational formalism, this procedure cannot be reduced to the first-order level if we take, for example, the particular situation $J^{2k-1}M = T^{2k-1}M$, then when $k = 1$ we have a Hamiltonian formalism on the tangent bundle, not an the cotangent bundle.

We will suppose that, at least for some Lagrangians, there is a Hamiltonian theory which can be developed over such manifold. If this is possible, then we will propose a way to related them in the next paragraph.

REMARK (6). There are other possible formulations. In a paper published in 1981, R. Arens proposed a method which consists in "reducing" the order of the Lagrangian by the introduction in the theory of a great number of variables, with the help of multipliers. In terms of Jet theory the new Lagrangian (of order 1) is defined on the manifold

$$J^1(N,J^1(J^k(N,M),\mathbb{R}).$$

Francaviglia & Krupka (1982) proposed a Hamiltonian

formalism on the manifold $J^k(J^{k-1}M)$ but this was erroneously based in a false theorem as it was shown by Dedecker (1984)(a), (1984)(b). But even in such formalism their Hamilton equations do not have the usual aspect, in the sense of the presence of all conjugate variables in the equations.

We develop now the first approach H(1), starting from the particular situation corresponding to the autonomous Generalized Particle Mechanics. We follow Aldaya & Azcárraga (1980) whose formalism was developed for higher order field theory.

Consider the trivial bundle $(\mathbb{R}\times E,p,\mathbb{R})$ and identify $J^k(\mathbb{R}\times E)$ with the manifold of k-jets of smooth mappings from $\mathbb{R}$ to E. The tangent bundle of order k, T^kE is the submanifold of $J^k(\mathbb{R},E)$ constructed from all smooth mappings from $\mathbb{R}$ to M with source at the origin of $\mathbb{R}$. The coordinate system for $(T^kE,\alpha^k,\mathbb{R})$ are represented by $(0,q_i,z^i_j)$, $1 \le i \le \dim E = n$, $1 \le j \le k$, where now we omitt the coefficients $(1/j!)$ for coordinates (we assume this only to simplify the expressions). Along a k-jet $\tilde{h}^k(0)$ one has $(q_i,q^i_{(j)})$, where - for technical reasons - we adopt the notation $q^k_{(j)}$ for the time derivative of order j of q^i, i.e., $q^i_{(j)} = q^{(j)i}$. We recall also that the canonical 1-forms on T^kE (see Example 1, §3.1) are locally expressed by

$$\theta^i_{(j-1)} = dz^i_{j-1} - z^i_j\, dt, \qquad 1 \le j \le k, \quad z^i_o = q^i$$

and, for a Lagrangian of order k, $L^k\colon T^kE \to \mathbb{R}$, the Poincaré-Cartan form is locally given by

$$\Omega_k = (\sum_{j=1}^{k} f_i^{(j)}\theta_{(j-1)}^i + L^k)dt.$$

Consider now the manifold of all 1-jets of non-constant smooth mappings from $\mathbb{R}$ to $T^{k-1}E$ with source at the origin of $\mathbb{R}$, $J_o^1(\mathbb{R},T^{k-1}E)$. Then $J_o^1(\mathbb{R},T^{k-1}E) = T(T^{k-1}E)$. A coordinate system is represented by

$$(q_i, z_s^i, Z_{,1}^i, Z_{s,1}^i), \qquad 1 \leq s \leq k-1.$$

If $g\colon \mathbb{R} \to T^{k-1}E$ is a smooth mapping such that $g(0) = \tilde{h}^{k-1}(0)$, for some $h\colon \mathbb{R} \to E$ then the local expression of $\tilde{g}^1(0)$ is

$$(q_i, q_{(s)}^i, Q_{,(1)}^i, Q_{(s),(1)}^i) \qquad (1 \leq s \leq k-1)$$

where

$$g^i(0) = (q_i, q_{(1)}^i, \ldots, q_{(k-1)}^i)$$

and

$$(d/dt)q_i(0) = Q_{,(1)}^i, \ldots, (d/dt)(q_{(s)}^i(0)) = Q_{(s),(1)}^i.$$

The dimension of $T(T^{k-1}E)$ is 2kn. Suppose now that

$$L^{1,k-1}\colon T(T^{k-1}E) \to \mathbb{R}$$

is a Lagrangian. Then from the canonical 1-forms on $T(T^{k-1}E)$,

$$\theta_{i;(s)}^{1,k-1} = dz_s^i - Z_{s,1}^i\, dt, \qquad 0 \leq s \leq k-1$$

and the Poincaré-Cartan form, one obtains from Hamilton's principle the set of equations

$$\frac{\partial L^{1,k-1}}{\partial z_s^i} - \frac{d}{dt}\left(\frac{\partial L^{1,k-1}}{\partial Z_{s,1}^i}\right) = 0 \qquad (25)_1$$

$$\begin{pmatrix} \frac{\partial}{\partial z^j_{,1}}\left(\frac{\partial L^{1,k-1}}{\partial z^i_{,1}}\right) \cdots \frac{\partial}{\partial z^j_{,1}}\left(\frac{\partial L^{1,k-1}}{\partial z^i_{k-1,1}}\right) \\ \vdots \qquad\qquad \vdots \\ \frac{\partial}{\partial z^j_{k-1,1}}\left(\frac{\partial L^{1,k-1}}{\partial z^i_{,1}}\right) \cdots \frac{\partial}{\partial z^j_{k-1,1}}\left(\frac{\partial L^{1,k-1}}{\partial z^i_{k-1,1}}\right) \end{pmatrix} \begin{pmatrix} \frac{dq^i}{dt} - z^i_{,1} \\ \vdots \qquad \vdots \\ \frac{dz^i_{k-1}}{dt} - z^i_{k-1,1} \end{pmatrix} = 0 \qquad (25)_2$$

When the 2nd system of equations $(25)_2$ admits only the trivial solution, the jet prolongation condition is fulfilled and the first system takes the Euler-Lagrange form

$$\frac{\partial L^{1,k-1}}{\partial q^i_{(s)}} - \frac{d}{dt}\left(\frac{\partial L^{1,k-1}}{\partial Q^i_{(s),(1)}}\right) = 0. \qquad (26)$$

Recalling that the Euler-Lagrange equations for $L^k: T^kE \to \mathbb{R}$ is

$$\frac{\partial L^k}{\partial q_i} + \sum_{\ell=1}^{k} (-1)\frac{d^\ell}{dt^\ell}\left(\frac{\partial L^k}{\partial q^i_{(\ell)}}\right) = 0. \qquad (27)$$

We have the following

PROPOSITION (4). Consider the above two Lagrangian functions. Then the equation (27) may be derived from equation (26) in terms of Lagrange multipliers.

Proof. We recall that every k-jet manifold can be regularly immersed into the jet prolongation of the corresponding (k-1) jet manifold. In our present situation this is obtained by $\varphi: T^kE \to T(T^{k-1}E)$ defined by

$$q_i(u(x)) = q_i(\varphi(u(x))),$$

$$z^i_1(u(x)) = Z^i_{,1}(\varphi(u(x))), \ldots, z^i_k(u(x)) = Z^i_{k-1,1}(\varphi(u(x))),$$

where $x \in E$ and $u \in Sec(T^kE)$.

A way to establish a relation between both Lagrangians is applying the multiplier rule:

$$L^{1,k-1} = L^k + \sum_{s=1}^{k-1} (-1)^s \lambda_i^s (q_{(s)}^i - Z_{s-1,1}^i) \qquad (28)$$

where, for simplicity, we have considered

$$L^{1,k-1} = L^{1,k-1}(q_i, q_{(r)}^i, Z_{,1}^i, Z_{r,1}^i), \qquad 1 \le r \le k-1$$

and L^k was identified with its expression induced by φ.

If we consider the expressions

$$\frac{\partial L^{1,k-1}}{\partial q_i} = \frac{d}{dt}\left(\frac{\partial L^{1,k-1}}{\partial Z_{,1}^i}\right); \qquad \frac{\partial L^{1,k-1}}{\partial q_{(r)}^i} = \frac{d}{dt}\left(\frac{\partial L^{1,k-1}}{\partial Z_{r,1}^i}\right),$$

then, taking into account (28), we get

$$\frac{\partial L^{1,k-1}}{\partial q_i} = \frac{d}{dt}\left(\frac{\partial L^k}{\partial Z_{,1}^i} + \lambda_i^1\right); \qquad \frac{\partial L^{1,k-1}}{\partial q_{(r)}^i} = \frac{d}{dt}\left(\frac{\partial L^k}{\partial Z_{r,1}} + (-1)^r \lambda_i^{r+1}\right).$$

On the other hand, deriving (28) with respect to $q_{(r)}^i$ gives

$$\frac{\partial L^{1,k-1}}{\partial q_{(r)}^i} = (-1)^r \lambda_i^r = \frac{d}{dt}\left(\frac{\partial L^{1,k-1}}{\partial Z_{r,1}^i}\right) = \frac{d}{dt}\left[\frac{\partial L^k}{\partial Z_{r,1}^i} + (-1)^r \lambda_i^{r+1}\right],$$

$$1 \le r \le k-1$$

and therefore

$$\frac{\partial L^{1,k-1}}{\partial q_i} = \frac{d}{dt}\left(\frac{\partial L^{1,k-1}}{\partial Z_{,1}^i}\right) = \frac{d}{dt}\left[\frac{\partial L^k}{\partial Z_{,1}^i} + \lambda_i^1\right] =$$

$$= \frac{d}{dt}\left[\frac{\partial L^k}{\partial Z_{,1}^i} - \frac{d}{dt}\left[\frac{\partial L^k}{\partial Z_{1,1}^i} - \lambda_i^2\right]\right] =$$

$$= \quad \dots \qquad \dots$$

$$= \frac{d}{dt}\left(\frac{\partial L^k}{\partial Z_{,1}^i}\right) - \frac{d^2}{dt^2}\left(\frac{\partial L^k}{\partial Z_{1,1}^i}\right) + \dots + (-1)^{k-1} \frac{d^{k-1}}{dt^{k-1}}\left(\frac{\partial L^k}{\partial Z_{k-1,1}^i}\right),$$

(recall that $\lambda^k \equiv 0$). As $(\partial L^{1,k-1}/\partial q_i) = (\partial L^k/\partial q_i)$ and from the above constrained relations we have the desired equations (27) when (26) is expressed along a k-jet prolongation.

□

We are now able to discuss the Hamiltonian formalism. Considering $L^{1,k-1}: T(T^{k-1}E) \to \mathbb{R}$ we define the (Legendre) transformation

$$\text{Leg}: T(T^{k-1}E) \to T^*(T^{k-1}E)$$

by

$$(q_i, z^i_r, Z^i_{,1}, Z^i_{r,1}) \longmapsto (q_i, z^i_r, P^{,1}_i, P^{r,1}_i)$$

where, for $v \in T(T^{k-1}E)$

$$P^{,1}_i(\text{Leg}(v)) = (\partial L^{1,k-1}/\partial Z^i_{,1})(v)$$

$$P^{r,1}_i(\text{Leg}(v)) = (\partial L^{1,k-1}/\partial Z^i_{r,1})(v), \qquad 1 \leq r \leq k-1$$

and, along jet prolongations we have

$$P^{,1}_i = \frac{\partial L^{1,k-1}}{\partial Q^i_{,(1)}} = \frac{\partial L^{1,k-1}}{\partial q^i_{(1)}}\ ;\ldots;\ P^{r,1}_i = \frac{\partial L^{1,k-1}}{\partial q^i_{(r+1)}} \qquad 1 \leq r \leq k-1 \tag{29}$$

(the reader may compare (29) with the momenta defined in Chapter II for the Lagrangian L^k). The Hamiltonian is now defined as a real smooth function $H^{1,k-1}$ on $T^*(T^{k-1}E)$ such that

$$H^{1,k-1} = \sum_{r=0}^{k-1} P^{r,1}_i\, Z^i_{r,1} - L^{1,k-1}$$

with the convention $P^{o,1}_i = P^{,1}_i$, etc. In terms of such $H^{1,k-1}$ one has

$$\Omega_k = \sum_{r=0}^{k-1} dP^{r,1}_i \wedge dZ^i_{r,1} - dH^{1,k-1} \wedge dt$$

which is the Poincaré-Cartan form associated to $H^{1,k-1}$. Hamilton equations take the form

$$\frac{\partial H^{1,k-1}}{\partial P_i^{r,1}} = \frac{dq^i_{(r+1)}}{dt} \; ; \qquad \frac{\partial H^{1,k-1}}{\partial q^i_{(r+1)}} = - \frac{dP_i^{r,1}}{dt} \tag{30}$$

along jet prolongations. Now, we have also seen in Chapter II that a Hamiltonian of order k (i.e., defined on $T^{2k-1}E$) is locally given by

$$H^k = \sum_{r=0}^{k-1} p_{i/r+1} \, q^i_{(r+1)} - L^k .$$

Let us determine the relation between both momenta. For this we consider again the immersion φ: for $v \in T^kE$ we have

$$(\varphi\circ \mathrm{Leg})^* \, q_i(v) = q_i(v)$$

$$(\varphi\circ \mathrm{Leg})^* Z^i_{,1}(v) = P_i^{,1}(v)$$

$$\vdots$$

$$(\varphi\circ \mathrm{Leg})^* Z^i_{r,1}(v) = P_i^{r,1}(v), \qquad 1 \le r \le k-1.$$

Now

$$P_i^{k-1,1} = \frac{\partial L^{1,k-1}}{\partial Z^i_{k-1,1}} = \frac{\partial L^k}{\partial Z^i_{k-1,1}} \; ;$$

$$P_i^{r,1} = \frac{\partial L^{1,k-1}}{\partial Z^i_{r,1}} = \frac{\partial L^k}{\partial Z^i_{r,1}} + \lambda_i^{r+1} , \qquad 1 \le r \le k-2.$$

Therefore, from an analogous calculation,

$$P_i^{r,1} = \sum_{s=0}^{k-r-1} (-1)^s \frac{d^s}{dt^s} \left(\frac{\partial L^k}{\partial q^i_{(s+r)}}\right) = p_{i/r+1} , \qquad 0 \le r \le k-1$$

along jet prolongation, and equatios (3) are the Hamiltonian form of equations (27).

For the general situation, that is, Ω^k is a Lagrangian density defined on the k-jet manifold J^kM, we have an analogous result (see Aldaya & Azcárraga (1980)). We embed J^kM into $J^1(J^{k-1}M)$ and then we consider

$$\mathfrak{L}^{1,k-1} = \mathfrak{L}^k + \sum_{s=1}^{k-1} (-1)^s \lambda_A^{a(s)} (z^A_{a(s)} - g^A_{a(s-1),a}).$$

The Hamiltonian form of the generating form is

$$d(\pi_A^{a(s),a}\, dz^A_{a(s)} \wedge \omega_a - \mathcal{H}^{1,k-1}\omega),$$

where

$$\pi_A^{a(s),a}(\mathrm{Leg}(v)) = \frac{\partial \mathfrak{L}^{1,k-1}}{\partial g^A_{a(s),a}}(v)$$

and $\mathcal{H}^{1,k-1}: J^{1*}(J^{k-1}M) \to \mathbb{R}$ is the corresponding Hamiltonian.

Let us now see the <u>second</u> point of view H(2). When there is only one independent variable we may develop the theory as usual, as it was shown in Chapter II. From the definition of the momenta we see that

$$p_{i/r} = p_{i/r}(q_j, q^j_{(1)}, \ldots, q^j_{(2k-r)}), \qquad 1 \leq r \leq k$$

and in particular

$$p_{i/k} = p_{i/k}(q_j, q^j_{(1)}, \ldots, q^j_{(k)}).$$

This function is a good candidate to be "canonically conjugate" to $q^i_{(k)}$. For this we impose that the matrix

$$\left(\frac{\partial p_{i/k}}{\partial q^i_{(k)}}\right)$$

be of maximal rank. The same may be assumed for the other matrices

$$\left(\frac{\partial p_{i/r}}{\partial q^j_{(2k-r)}}\right), \qquad r < k,$$

and from these regularity conditions we define the Legendre

transformation locally by

$$(q_j, q^j_{(1)}, \dots, q^j_{(k)}, \dots, q^j_{(2k-1)}) \mapsto (q_j, q^j_{(1)}, \dots, q^j_{(k-1)}, p_{i/k}, \dots, p_{i/1})$$

from $T^{2k-1}E$ to itself. The Hamiltonian formalism now follows directly as in the standard case. As seen before, this procedure cannot be canonically extended for the general situation. But if it is possible to express the Hamiltonian form of the generalized Euler-Lagrange equations then we need to "reduce" the domain of definition for the Hamiltonian of order k. For example, if we take $k = 3$, in J^5M the coordinates are $(x_a, y^A, z^A_{a_1}, \dots, z^A_{a_1 \dots a_5})$. Now the momenta are defined by

$$p_{A/a_1} = \frac{\partial \mathfrak{L}^3}{\partial z^A_{a_1}} - \frac{d}{dx_{a_2}}\left(\frac{\partial \mathfrak{L}^3}{\partial z^A_{a_1 a_2}}\right) + \frac{d^2}{dx_{a_2}\, dx_{a_3}}\left(\frac{\partial \mathfrak{L}^3}{\partial z^A_{a_1 a_2 a_3}}\right)$$

$$p_{A/a_1 a_2} = \frac{\partial \mathfrak{L}^3}{\partial z^A_{a_1 a_2}} - \frac{d}{dx_{a_3}}\left(\frac{\partial \mathfrak{L}^3}{\partial z^A_{a_1 a_2 a_3}}\right)$$

$$p_{A/a_1 a_2 a_3} = \frac{\partial \mathfrak{L}^3}{\partial z^A_{a_1 a_2 a_3}}$$

where $\mathfrak{L}^3 : T^3M \to \mathbb{R}$ is the Lagrangian density. Proceeding as above we see that $p_{A/a_1 a_2 a_3}$ may be conjugate to $z^A_{a_1 a_2 a_3}$. The regularity condition for the matrices

$$\left(\frac{\partial p_{A/a(r)}}{\partial z^A_{a(6-r)}}\right), \qquad 1 \le r \le 2$$

does not assure that the corresponding z's are conjugate to p_{A/a_1} and $p_{A/a_1 a_2}$ since they are not in the same number. We may solve this problem by considering a submanifold of

$J^{2k-1}M$, with a certain co-dimension, for which the Hamiltonian will be defined. From the regularity of

$$\left(\frac{\partial p_{A/a(r)}}{\partial z^B_{a(2k-r)}}\right)$$

when $r = k$, we may use local coordinates from $J^{k-1}M$ together with the momenta $p_{A/a(k)}$ as a local coordinate system for J^kM. When $r = k-1$, there are a certain number of coordinates (represented symbolically by $z_A^{b(k+1)}$) such that

$$(x_a, y^A, \ldots, z^A_{a(k-1)}, \ldots, p_{A/a(k)}, p_{A/a(k-1)}, z_A^{b(k+1)})$$

is a coordinate system for $J^{k+1}M$. If we take the holonomic prolongation up to order $2k-1$ (inclusive) we obtain a coordinate system of type

$$(x_a, y^A, \ldots, z^A_{a(k-1)}, p_{A/a(k)}, \ldots, p_{A/a(1)}, z_A^{b(2k-1)})$$

and we define $\hat{J}^{2k-1}M$ as the submanifold of $J^{2k-1}M$ characterized by the above coordinates without $z_A^{b(2k-1)}$. The Hamiltonian in such context will be a function $\mathcal{H}: \hat{J}^{2k-1}M \to \mathbb{R}$ such that

$$\mathcal{H} = \sum_{\ell=1}^{k} p_{A/a(\ell)} z^A_{a(\ell)} - \mathcal{L}(x_a, y^A, \ldots, z^A_{a(k)}).$$

This reasoning is inspired in Shadwick (1982) for the situation where $a_1 \leq a_2 \leq \ldots \leq a_k$ in the definition of G_k. Shadwick (see Exercise (1)) considered the case where the symmetry of the indexes is included and therefore the definition of the p's must be modified (for example, when $k = 2$, $p_{A/a_1a_2} = (1/2)\, \partial\mathcal{L}/\partial y^A_{a_1a_2}$). The general situation is of course more complicated. In our present formulation we may show analytically the equivalence of Lagrange and Hamilton equations as it was done in Chapter II, but this does not assure, in general, that there is an injective and surjective mapping between the set of extremals of the variational problem generated by the "Poincaré-Cartan" form on $J^{2k-1}M$.

3.5. A possible relation between two Hamiltonian formalisms.

Let us consider a Lagrangian L of type $L = L(x,y^A,z^A_{(1)},z^A_{(2)})$, where $z^A_{(i)} = d^i y^A/dx^i$, $1 \le i \le 2$. Following the Jacobi-Ostrogradsky method we may put the Lagrangian equations in the canonical form by considering

$$q^A_1 = y^A, \quad q^A_2 = z^A_{(1)}, \quad p_{A/1} = \frac{\partial L}{\partial z^A_{(1)}} - \frac{d}{dx}\left(\frac{\partial L}{\partial z^A_{(2)}}\right), \quad p_{A/2} = \frac{\partial L}{\partial z^A_{(2)}}$$

The Hamiltonian is then defined by

$$H = p_{A/1}q^A_2 + p_{A/2}z^A_{(2)} - L(x,q^A_1,q^A_2,z^A_{(2)}).$$

If we suppose that H is a function depending only on $(x,q^A_1,q^A_2,p_{A/1},p_{A/2})$ then the value $z^A_{(2)}$ has to be eliminated through equation

$$p_{A/2} = \partial L/\partial z^A_{(2)} .$$

An exception occurs when L is degenerate. For example, Hayes (1969) considered the one dimensional Lagrangian defined by

$$L = (-1/2)mz_{(2)}-V(y), \quad (V = \text{potential energy;} \quad m = \text{mass})$$

with Hamiltonian

$$H = p_1q_2 + p_2z_{(2)} + (1/2)mq_1z_{(2)} + V(q_1)$$

where $q_1 = y$, $q_2 = z_{(1)} = dy/dx$, $p_1 = (1/2)mz_{(1)}$, $p_2 =$ $= (-1/2)my$. Therefore we cannot eliminate $z_{(2)}$. If we replace the q's and p's by their values then we eliminate $z_{(2)}$. However the canonical equations are not equivalent to the Lagrangian ones. Hayes suggested to re-write the Hamiltonian by replacing $z_{(2)}$ by $(-1/m)(\partial V/\partial y)$ and taking into account the expressions for p_1 and p_2. The Hamiltonian takes the following form:

$$H = (p_1^2/m) + (mq_2^2/4) + (V(q_1)/2) + V(-(2/m)p_2/2)$$

and the equivalence is established. But the choice of such Hamiltonian is not unique as it was shown by Ryan (1972). If we consider the transformation

$$Q = (-2/m)p_2 \ ; \qquad P = (1/2)mq_2$$

then

$$H = (p_1^2/m) + (P^2/m) + (1/2)V(q_1) + (1/2)V(Q).$$

This Hamiltonian gives the same canonical equations for

$$H(q_1,p_1) \overset{\text{def}}{=} (p_1^2/2m) + V(q).$$

To solve the unicity problem Ryan suggested to consider equivalent Lagrangians of type $L = L(x,y^A,z^A_{(1)},z^A_{(2)}) =$ $= z^A_{(2)}\ F(x,y^A,z^A_{(1)}) + G(x,y^A,z^A_{(1)})$. This method works only for the one-dimensional situation as it was shown by Kimura (1972). Such Lagrangians are degenerate and Kimura showed that if we employ the so-called Dirac theory of constraints then we obtain a consistent Hamiltonian formalism without using any kind of equivalent Lagrangians.

We want to establish a similar result for the formalisms

$H(1)$ and $H(2)$. As we have seen, for a Lagrangian of order k we cannot define directly the corresponding Hamiltonian of order k on $J^k M$. Our idea may be summarized in the following scheme:

$$R \xleftarrow{\quad \mathcal{L}^k \quad} J^k M \xleftarrow{\quad \rho_k^{2k-1} \quad} J^{2k-1} M \xrightarrow{\quad \text{embed.} \quad} J^1(J^{2k-2}M) \longrightarrow J^{1*}(J^{2k-2}M)$$

$$\Big\downarrow (\pi_k^{2k-1})^* \mathcal{L}^k \qquad \hat{J}^{2k-1} M \qquad H^{2k-1} \Big/ H^{1,2k-2}$$

$$\mathbb{R} \qquad\qquad\qquad\qquad \mathbb{R}$$

$$? $$

                                        constraint Dirac theory

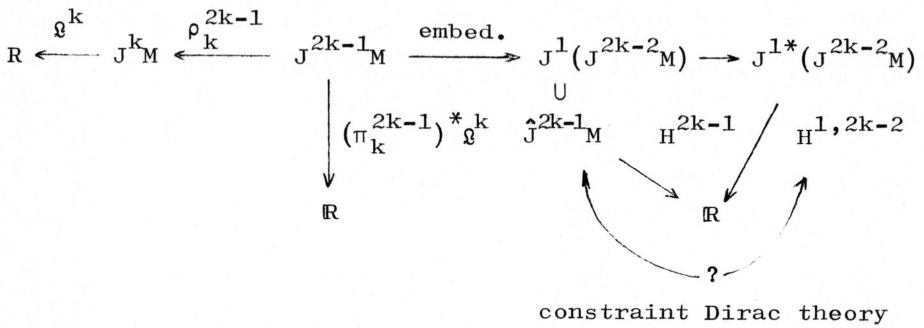

that is,

THEOREM. The Hamiltonian H^{2k-1} formalism can be viewed locally as a standard theory of constraints in the Dirac sense for the Hamiltonian $H^{1,2k-2}$ formalism on the dual manifold of $J^1(J^{2k-2}M)$.

We are assuming that there is such H^{2k-1} formalism and it is also clear that the local character of the theory is implicit in the theorem. In what follows we suppose that indices are ordered $a_1 \leq a_2 \leq \ldots$. We refer the reader to Appendix B for a discussion about Dirac theory.

For sake of brevity let us consider $k = 2$. The following manifolds and functions will be used: $J^2 M$, $J^3 M$, $J^1(J^2 M)$, $J^{1*}(J^2 M)$ and

$$\mathcal{L}^2 : J^2 M \to \mathbb{R}; \qquad \mathcal{L}^2 = \mathcal{L}^2(x_a, y^A, z_a^A, z_{ab}^A)$$

$$\mathcal{L}^3 : J^3 M \to \mathbb{R}; \qquad \mathcal{L}^3 = \mathcal{L}^3(x_a, y^A, z_a^A, z_{ab}^A, z_{abc}^A)$$

$$\mathcal{L}^{1,2} : J^1(J^2 M) \to \mathbb{R}; \qquad \mathcal{L}^{1,2} = \mathcal{L}^{1,2}(x_a, y^A, z_a^A, z_{ab}^A, g_{,a}^A, g_{b,a}^A, g_{bc,a}^A)$$

$$\hat{\mathcal{H}}^2 : J^3M \to \mathbb{R}; \quad \hat{\mathcal{H}}^2 = \hat{\mathcal{H}}^2(x_a, y^A, z^A_a, p_{A/a}, p_{A/ab}) = \mathcal{H}^3$$

$$\mathcal{H}^{1,2} : J^{1*}(J^2M) \to \mathbb{R}; \quad \mathcal{H}^{1,2} = \mathcal{H}^{1,2}(x_a, y^A, z^A_{ab}, \pi^a_A, \pi^{ab}_A, \pi^{abc}_A)$$

Let us first consider the Lagrangian $\mathfrak{L}^{1,2}$. For us this is a function of order 1 since we may "forget" that J^2M is a jet manifold. For order-one Lagrangians the field equations have (symbolically) the form

$$\frac{\partial \mathfrak{L}^{1,2}}{\partial u} - \frac{d}{dx_a}\left(\frac{\partial \mathfrak{L}^{1,2}}{\partial u_a}\right) = 0. \tag{31}$$

Now, correctly replacing the variables involved:

$$\text{if} \quad u = z^A_{bc} \quad \text{then} \quad u_a = g^A_{bc,a}$$

$$\text{if} \quad u = z^A_b \quad \text{then} \quad u_a = g^A_{b,a}, \quad \text{etc}$$

and developing (31), we have:

(i) $$\frac{\partial^2 \mathfrak{L}^{1,2}}{\partial g^A_{,a} \partial g^B_{,\alpha}}\left(\frac{\partial g^B_{,\alpha}}{\partial x_a}\right) = \frac{\partial \mathfrak{L}^{1,2}}{\partial y^A} - \frac{\partial^2 \mathfrak{L}^{1,2}}{\partial g^A_{,a}\partial x_a} - \frac{\partial^2 \mathfrak{L}^{1,2}}{\partial g^A_{,a}\partial x_a} - \frac{\partial^2 \mathfrak{L}^{1,2}}{\partial g^A_{,a}\partial y^B}\frac{\partial y^B}{\partial x_a} - \cdots - \frac{\partial^2 \mathfrak{L}^{1,2}}{\partial g^A_{,a}\partial g^B_{\beta\gamma,\alpha}}\frac{\partial g^B_{\beta\gamma,\alpha}}{\partial x_a},$$

(ii) $$\frac{\partial^2 \mathfrak{L}^{1,2}}{\partial g^A_{b,a} \partial g^B_{\beta,\alpha}}\left(\frac{\partial g^B_{\beta,\alpha}}{\partial x_a}\right) = \frac{\partial \mathfrak{L}^{1,2}}{\partial z^A_a} - \frac{\partial^2 \mathfrak{L}^{1,2}}{\partial g^A_{b,a}\partial x_a} - \frac{\partial^2 \mathfrak{L}^{1,2}}{\partial g^A_{b,a}\partial y^B}\frac{\partial y^B}{\partial x_a} - \cdots - \frac{\partial^2 \mathfrak{L}^{1,2}}{\partial g^A_{b,a}\partial g^B_{\beta\gamma,\alpha}}\frac{\partial g^B_{\beta\gamma,\alpha}}{\partial x_a},$$

(iii) $$\frac{\partial^2 \mathfrak{L}^{1,2}}{\partial g^A_{bc,a} \partial g^B_{\beta\gamma,\alpha}}\left(\frac{\partial g^B_{\beta\gamma,\alpha}}{\partial x_a}\right) = \frac{\partial \mathfrak{L}^{1,2}}{\partial z^A_{ab}} - \frac{\partial^2 \mathfrak{L}^{1,2}}{\partial g^A_{bc,a}\partial x_a} - \frac{\partial^2 \mathfrak{L}^{1,2}}{\partial g^A_{bc,a}\partial y^B}\frac{\partial y^B}{\partial x_a} - \cdots - \frac{\partial^2 \mathfrak{L}^{1,2}}{\partial g^A_{bc,a}\partial g^B_{,\alpha}}\frac{\partial g^B_{,\alpha}}{\partial x_a} - \frac{\partial^2 \mathfrak{L}^{1,2}}{\partial g^A_{bc,a}\partial g^B_{\beta,\alpha}}\frac{\partial g^B_{\beta,\alpha}}{\partial x_a}.$$

The variables $g^A_{,a}$, $g^A_{b,a}$ etc play the role of standard velocities and so the first term in the left hand side of (i)-(iii) gives us the condition for $\mathfrak{L}^{1,2}$ to be or not degenerate. Putting

$$\pi_A^{bc,a} = \frac{\partial \mathfrak{L}^{1,2}}{\partial g^A_{bc,a}}, \quad \pi_A^{b,a} = \frac{\partial \mathfrak{L}^{1,2}}{\partial g^A_{b,a}} \quad \text{and} \quad \pi_A^{a} = \frac{\partial \mathfrak{L}^{1,2}}{\partial g^A_{,a}}$$

we may obtain a system of equations as in Proposition $(3)_1$; one set is

$$\begin{pmatrix} \dfrac{\partial \pi_A^{,a}}{\partial g^B_{,\alpha}} & \dfrac{\partial \pi_A^{b,a}}{\partial g^B_{,\alpha}} & \dfrac{\partial \pi_A^{bc,a}}{\partial g^B_{,\alpha}} \\[2ex] \dfrac{\partial \pi_A^{,a}}{\partial g^B_{\beta,\alpha}} & \dfrac{\partial \pi_A^{b,a}}{\partial g^B_{\beta,\alpha}} & \dfrac{\partial \pi_A^{bc,a}}{\partial g^B_{\beta,\alpha}} \\[2ex] \dfrac{\partial \pi_A^{,a}}{\partial g^B_{\beta\gamma,\alpha}} & \dfrac{\partial \pi_A^{b,a}}{\partial g^B_{\beta\gamma,\alpha}} & \dfrac{\partial \pi_A^{bc,a}}{\partial g^B_{\beta\gamma,\alpha}} \end{pmatrix} \begin{pmatrix} \dfrac{\partial y^A}{\partial x_a} - g^A_{,a} \\[2ex] \dfrac{\partial z^A_b}{\partial x_a} - g^A_{b,a} \\[2ex] \dfrac{\partial z^A_{bc}}{\partial x_a} - g^A_{bc,a} \end{pmatrix} = 0.$$

We will say that a <u>degenerate</u> situation occurs if the jet prolongation condition is not performed, i.e., the above system has non-trivial solution. We will assume, for simplicity, that all terms in the last column are zero ($\pi_A^{bc,a}$ depends only on x_a, y^A and so they have no place in this theory). We suppose also that the other blocks are of maximal rank.

As we have seen before, equations (31) (expressed in terms of the variables of $J^1(J^2M)$) can be related to equations

$$\frac{\partial \mathfrak{L}^3}{\partial y^A} - \frac{d}{dx_a}\left(\frac{\partial \mathfrak{L}^3}{\partial z^A_a}\right) + \ldots - \frac{d^3}{dx_a dx_b dx_c}\left(\frac{\partial \mathfrak{L}^3}{\partial z^A_{abc}}\right) = 0 \qquad (32)$$

where $\mathfrak{L}^3 : J^3M \to \mathbb{R}$ is a Lagrangian related to $\mathfrak{L}^{1,2}$ in terms of multipliers. For such $\mathfrak{L}^3$ we consider the induced variational problem for the form

$$p_{A/a}\theta^A \wedge \omega_a + p_{A/ab}\theta^A_a \wedge \omega_b + p_{A/abc}\theta^A_{ab} \wedge \omega_c + \mathfrak{L}^3\omega$$

and we obtain a system of equations where the system

$$\begin{pmatrix} \dfrac{\partial p_{A/a}}{\partial y^B} & \dfrac{\partial p_{A/ab}}{\partial y^B} & \dfrac{\partial p_{A/abc}}{\partial y^B} \\[2ex] \dfrac{\partial p_{A/a}}{\partial z^B_{\alpha\beta\gamma}} & \dfrac{\partial p_{A/ab}}{\partial z^B_{\alpha\beta\gamma}} & \dfrac{\partial p_{A/abc}}{\partial z^B_{\alpha\beta\gamma}} \\[2ex] \dfrac{\partial p_{A/a}}{\partial z^B_{\alpha\beta\gamma\theta}} & \dfrac{\partial p_{A/ab}}{\partial z^B_{\alpha\beta\gamma\theta}} & 0 \\[2ex] \dfrac{\partial p_{A/a}}{\partial z^B_{\alpha\beta\gamma\theta\mu}} & 0 & 0 \end{pmatrix} \begin{pmatrix} \dfrac{\partial y^A}{\partial x_a} - z^A_a \\[2ex] \dfrac{\partial z^A_b}{\partial x_a} - z^A_{ab} \\[2ex] \dfrac{\partial z^A_{ab}}{\partial x_a} - z^A_{abc} \end{pmatrix} = 0$$

Gives also the conditions for regularity. The functions p's (defined in page 196) are the momenta and along jet prolongations they are the π's. In particular $p_{A/abc} = \pi_A^{bc,a} = 0$ and therefore equations (32) are of order 4 and not 6. We may say that we are in the presence of a theory involving Lagrangians of order 2 and not 3. A way to obtain such Lagrangian defined on J^3M is given by the canonical projection $\rho^3_2 : J^3M \to J^2M$, $(x_a, y^A, z^A_a, z^A_{ab}, z^A_{abc}) \longmapsto (x_a, y^A, z^A_a, z^A_{ab})$. We put

$$(\rho^3_2)^*\mathfrak{L}^2 = \hat{\mathfrak{L}}^2(x_a, y^A, z^A_a, z^A_{ab}, \hat{z}^A_{abc}),$$

where $\wedge$ over a term indicates that it should be omitted.

This is our $\mathfrak{L}^3$ and the Hamiltonian counterpart is a function $\hat{H}^2$ (or H^3) defined by

$$\hat{\mathcal{H}}^2 = p_{A/a} z^A_a + p_{A/ab} z^A_{ab} - \hat{\mathfrak{L}}^2(x_a, y^A, z^A_a, z^A_{ab}, \hat{z}^A_{abc})$$

where now

$$p_{A/a} = \frac{\partial \mathfrak{L}^2}{\partial z^A_a} - \frac{d}{dx_b}\left(\frac{\partial \mathfrak{L}^2}{\partial z^A_{ab}}\right); \qquad p_{A/ab} = \frac{\partial \mathfrak{L}^2}{\partial z^A_{ab}} .$$

The function $\hat{\mathcal{H}}^2 = \hat{\mathcal{H}}^2(x_a, y^A, z^A_a, p_{A/a}, p_{A/ab})$ is defined on a submanifold $\hat{J}^3$ of J^3M regularly immersed into $J^{1*}(J^2M)$.

From the degeneracy assumption on $\mathfrak{L}^{1,2}$ the Hamiltonian $\mathcal{H}^{1,2}$: $J^{1*}(J^2M) \to R$ is <u>associated</u> to the above Lagrangian $\hat{\mathfrak{L}}^2$: $J^2M \to R$ and we may consider $\mathcal{H}^{1,2}$ as being $\hat{\mathcal{H}}^2$ and if we represent by $\bar{\mathcal{H}}^{1,2}$ any extension of $\mathcal{H}^{1,2}$ to an open set U which characterizes locally $\hat{J}^3M$, we have that the Hamiltonian must be of the form

$$h^{1,2} = \bar{\mathcal{H}}^{1,2} + u^a_A f^A_a$$

where the f^A_a are the functions which locally define $\hat{J}^3M$ and u^a_A are multipliers. It is now clear that

$$h^{1,2} = \pi^{,a}_A g^A_{,a} + \pi^{b,a}_A g^A_{b,a} - \hat{\mathfrak{L}}^2 + u^a_A(z^A_a - g^A_{,a})$$

$$= (\pi^{,a}_A - u^a_A) g^A_{,a} + \pi^{b,a}_A g^A_{b,a} + u^a_A z^A_a - \hat{\mathfrak{L}}^2 , \tag{33}$$

since $\mathcal{H}^{1,2} = \pi^{,a}_A g^A_a + \ldots - \mathfrak{L}^{1,2}$ and $\mathfrak{L}^{1,2} = \hat{\mathfrak{L}}^2 - u^a_A(z^A_a - g^A_{,a})$.

Now

$$u^a_A = -\frac{d}{dx_b}\left(\frac{\partial \hat{\mathfrak{L}}^2}{\partial z^A_{ab}}\right) + \frac{d^2}{dx_b dx_c}\frac{\partial \hat{\mathfrak{L}}^2}{\partial z^A_{abc}}$$

$$\pi^{,a}_A = \frac{\partial \hat{\mathfrak{L}}^2}{\partial g^A_{,a}} + u^a_A ; \qquad \pi^{b,a}_A = \frac{\partial \hat{\mathfrak{L}}^2}{\partial g^A_{b,a}}$$

and (33) takes the form

$$h^{1,2} = \frac{\partial \hat{\mathfrak{L}}^2}{\partial g^A_{,a}} g^A_{,a} + \left(- \frac{d}{dx_b} \frac{\partial \hat{\mathfrak{L}}^2}{\partial z^A_{ab}} + \frac{d^2}{dx_b dx_c} \frac{\partial \hat{\mathfrak{L}}^2}{\partial z^A_{abc}}\right) z^A_a +$$

$$+ \frac{\partial \hat{\mathfrak{L}}^2}{\partial g^A_{b,a}} g^A_{b,a} - \hat{\mathfrak{L}}^2 .$$

As along $\hat{J}^3 M, g^A_{,a} = z^A_{,a}$ we have effectivelly that $h^{1,2}$ is the corresponding Hamiltonian of order 2 associated to $\mathfrak{L}^2$;

$$\hat{\mathfrak{L}}^2 = \mathfrak{L}^2(x_a, y^A, z^A_a, z^A_{ab}, \hat{z}^A_{abc})$$

and

$$\mathcal{H}^2 \overset{\text{def}}{=} h^{1,2}/\hat{J}^3 M = p_{A/a} z^A_a + p_{A/ab} z^A_{ab} - \mathfrak{L}^2 .$$

3.6. Some comments on Noether's theorem in the generalized formalism.

Noether's theorem states that a conserved quantity is determined when a (infinitesimal) transformation leaves the action integral invariant. For regular Lagrangians these conserved quantities may be obtained from the integral defined by the Poincaré-Cartan form since first order variational problems are equivalent to variational problems of order zero. In the case of higher order variational problems involving many independent variables this equivalence is not canonically obtained as we have remarked before, but we may develop an alternative formalism if we embed $J^k M$ into $J^1(J^{k-1}M)$, (cf. Aldaya & Azcárraga (1978), (1980)). Information about a direct treatement on $J^k M$ may be found in Garcia & Muñoz (1983) and Muñoz (1983).

Let us assume $k = 1$, $J^1M = R\times TE$, i.e., the fibered manifold (M,p,N) is trivially fibered by $M = R\times E$, $N = R$. Let $L: \mathbb{R}\times TE \to \mathbb{R}$ be a regular Lagrangian and $\Omega_H = \omega - dH \wedge dt$, the generating form obtained by exterior differentiation of the Poincaré-Cartan form associated with $H: \mathbb{R}\times T^*E \to \mathbb{R}$,

$$\theta_H = \lambda - Hdt = p_i dq_i - Hdt,$$

where λ is the Liouville form canonically defined on T^*E (and so $\omega = -d\lambda$). θ_H is obtained from the Poincaré-Cartan form associated with the (regular) Lagrangian L,

$$\theta_L = p_i\theta_i - L\,dt = p_i dq_i - (p_i v_i + L)dt,$$

where $\theta_i = dq_i - v_i dt$.

For each time $t \in \mathbb{R}$, consider the Hamiltonian $h = H/t \times T^*E$ and the corresponding associated vector field X_h on T^*E, which verifies the symplectic equation

$$i_{X_h}\omega = dh.$$

Suppose that Y is a vector field on T^*E such that the Lie derivative

$$\mathcal{L}_Y\lambda = 0.$$

Then

$$i_Y d\lambda = -i_Y\omega = -d(i_Y\lambda)$$

and

$$i_Y\omega(X_h) = d(i_Y\lambda)(X_h) \Rightarrow$$

$$\Rightarrow i_{X_h}\omega(Y) = dh(Y) = d(i_Y\lambda)(X_h).$$

The 1-form $d(i_Y\lambda)$ generates a vector field Z_Y (depending on Y) such that

$$i_{Z_Y}\omega = d(i_Y\lambda).$$

Therefore

$$dh(Y) = i_{Z_Y}\omega(X_h) = \omega(Z_Y, X_h) = \{i_Y\lambda, h\},$$

where $\{\ ,\ \}$ is the Poisson brackets.

We recall here that a function is a <u>first integral</u> of a differential equation if it is constant along the integral curves (solutions) of the equation. If $\varphi(t)$ is an integral curve of the vector field X and if f is a first integral of X then $(d/dt)(f(\varphi(t)))) = X(f(\varphi(t)))) = 0$, i.e., f is constant along $\varphi(t)$. We extend such definition to forms by saying that a 1-form σ is first integral fo X if $\sigma(X) = 0$ (when $\sigma = df$, $i_X df = df(X) = X(f) = 0$).

Suppose that $F: T^*E \to \mathbb{R}$ is a smooth function (depending implicity on the time t). If we take the total derivative of F with respect to t and if we take into account Hamilton equations for h then the restriction of dF/dt to the integrals of the system gives

$$(dF/dt) = (\partial F/\partial dq_i)(\partial h/\partial p_i) - (\partial F/\partial p_i)(\partial h/\partial q_i) = \{F, h\},$$

and F is a constant of motion if and only if $[F, h] = 0$ (for the non-autonomous situation we have $dF/dt = \partial F/\partial t + [F, h]$). From the first expression for the Poisson bracket it follows that: "dh is a first integral of Y if and only if the function $i_Y\lambda$ (or the form $d(i_Y\lambda)$) is a first integral (a constant of motion) of the canonical equations determined by the Hamiltonian h (or by the vector field X_h)". We recognize in such result a version of Noether's theorem.

The extension of the above result to the non-autonomous situation is straightforward. The Hamiltonian vector field X_H associated to $H: \mathbb{R}\times T^*E \to \mathbb{R}$ is locally given by

$$X_H = \frac{\partial}{\partial t} + \left(\frac{\partial H}{\partial p_i}\frac{\partial}{\partial q_i} - \frac{\partial H}{\partial q_i}\frac{\partial}{\partial p_i}\right) = \frac{\partial}{\partial t} + X_h \, .$$

Clearly, if Y is a vector field on $\mathbb{R}\times T^*E$ such that $Y(t) = 1$ and $i(Y)\Omega_H = 0$ then $Y = X_H$. Also, for any vector field Z on $\mathbb{R} \times T^*E$ it can be shown by a direct calculation (cf. Aldaya & Azcárraga (1980), for example) that along sections $s \in Sec(\mathbb{R}\times T^*E)$ verifying the corresponding Hamilton equations one has $s^*(i_Z\Omega_H) = 0$. It can equally be shown that for a smooth function F on $\mathbb{R} \times T^*E$ the following equality holds

$$\mathcal{L}_{X_H} F = \partial F/\partial t + \{F,H\} = dF/dt.$$

We have considered above a vector field Y on T^*E. This vector field must be related to some transformation on E since the perturbations act on curves in such manifold. In physics, the symmetry properties are characterized by transformation which leave the Hamiltonian (or the Lagrangian) conserved under their action. The constants of motion are generating functions of (infinitesimal) canonical transformations (the term infinitesimal means small perturbation on the coordinates). Geometrically this can be characterized by vector fields for which the corresponding flow leaves invariant the dynamical (field) equations. Therefore, in the non-autonomous situation (or in the autonomous ones) we say that a vector field X on $\mathbb{R}\times E$ generates a symmetry property if for the lifted vector field Y on $\mathbb{R} \times T^*E$ one has $\mathcal{L}_Y\theta_H = 0$

(a more general situation may be obtained if we say that there is a function F depending on Y such that $\mathcal{L}_Y\theta_H = dF$). The vector field Y is $\bar{X}^1$, the projectable vector field along X, i.e., in local coordinates

$$\bar{X}^1 = X + Y_i\ \partial/\partial p_i = \partial/\partial t + X_i\ \partial/\partial q_i + Y_i\ \partial/\partial p_i .$$

If we fix the time, we have $Y_i = p_i(\partial X_i/\partial q_i)$ and $\bar{X}^1$ verifies along the extremals the symplectic equation for the Hamiltonian $i_Y\lambda$.

Developing $\mathcal{L}_{\bar{X}^1}\ \theta_H = 0$, we have

$$i_{\bar{X}^1}\ d\theta_H + d\ i_{\bar{X}^1}\theta_H = i_{\bar{X}^1}\ \Omega_H + d\ i_{\bar{X}^1}\ \theta_H = 0.$$

Now, as we are interested on first integrals of the Hamiltonian equations, $i_{\bar{X}^1}\ \Omega_H = 0$ along sections which are extremals of the (zero order) variational problem, we obtain the Noether invariant

$$d\ i_{\bar{X}^1}\ \theta_H = d(p_i dq_i(\bar{X}^1) - Hdt(\bar{X}^1))$$

$$= d(p_i \frac{\partial X_i}{\partial q_i} - H) = 0$$

(in the general situation we have $d\ i_{\bar{X}^1}\ \theta_H - F = 0$). If we develop $\mathcal{L}_{\bar{X}^1}\ \theta_H$ in the following two ways,

$$\mathcal{L}_{\bar{X}^1}\ \theta_H = \mathcal{L}_{\bar{X}^1}\ \lambda - \mathcal{L}_{\bar{X}^1}\ Hdt =$$

$$= \mathcal{L}_{\bar{X}^1}\ \lambda - i_{\bar{X}^1}\ dH \wedge dt - di_{\bar{X}^1}\ H\ dt,$$

$$\mathcal{L}_{\bar{X}^1}\ \theta_H = i_{\bar{X}^1}\ \theta_H + di_{\bar{X}^1}\ \theta_H$$

$$= i_{\bar{X}^1}\ d\lambda - i_{\bar{X}^1}\ dH \wedge dt + di_{\bar{X}^1}\ \theta_H =$$

$$= i_{\bar{X}^1} d\lambda - i_{\bar{X}^1} dH \wedge dt + di_{\bar{X}^1} d\lambda + di_{\bar{X}^1} Hdt,$$

we obtain the identity $\mathcal{L}_{\bar{X}^1} \lambda = i_{\bar{X}^1} d\lambda + di_{\bar{X}^1} \lambda$ and therefore $\mathcal{L}_{\bar{X}^1} \theta_H = 0$ implies $i_{\bar{X}^1} d\lambda = -di_{\bar{X}^1} d\lambda$, the equality examinated before.

A first extension of the above procedure concerns multiple integrals variational problems of order one, i.e., we consider a Lagrangian $L: J^1M \to \mathbb{R}$, where now $\dim N > 1$. Taking into account the regular conditions on the characteristic matrix of L (cf. §3.3) we may associate the extremals of the first order variational problem generated by L with the zeroth problem for the Poincaré-Cartan form

$$G_H = (\partial L/\partial z^A_a) dy^A \wedge \omega_a - H\omega.$$

For the higher order situation we may adopt the embedding procedure of J^kM into $J^1(J^{k-1}M)$ and an analogous study may be developed for such point of view (for more details see Aldaya & de Azcárraga (1980)(b)).

We may also examine Noether invariants directly for a k-th order variational problem. We define an infinitesimal symmetry Y on M by saying that $\mathcal{L}_{\bar{Y}^k} (\mathfrak{L}\omega) = 0$, where $\bar{Y}^k$ is the unique infinitesimal contact transformation of order k projectable on Y. In such a case, for each critical section, we have the Noether invariant $(\tilde{s}^{2k-1})^* \, di_{\bar{Y}^{2k-1}} \Omega$, where Ω is the Poincaré-Cartan form. The global study of Noether invariants may be founded in the article of Muñoz (1983).

The higher order situation may be viewed also in terms of the traditional local procedure. For example it suffices to consider formula $(9)_1$ in §2.3 to get along the extremals

$$\sum_{r=0}^{k-1} \frac{d}{dx_{a_{r+1}}} \left[\frac{\delta J}{\delta y^A_{a(r+1)}} f^A_{a(r)} - \left(\frac{\delta J}{\delta y^A_{a(r+1)}} y^A_{a(r)a_{r+1}} - \mathfrak{L} \right) g^A_{a_{r+1}} \right] = 0,$$

(this formula may be extended to include the generalized momentum tensor defined in §2.6 (see Musicki (1978) and also Anderson (1973) for $\dim N = 1$).

It is also interesting to study the conditions for the inversion of Noether's theorem, i.e., "to every conserved quantity there corresponds an i.c.t. for which the variation vanishes". For a local study we refer to the articles of Palmieri & Vitale (1970); Candotti, Palmiere & Vitale (1970); Candotti, Palmieri & Vitale (1972)(a); Candotti, Palmieri & Vitale (1972)(b); Rosen (1971), (1972). For a detailed analysis and bibliography from the classical point of view see Hill (1951) and Denman (1965).

III.4 - Further considerations and some examples

4.1. Lagrangians with higher partial derivatives of different orders.

We have so far examined Lagrangians $\mathfrak{L}$ of type

$$\mathfrak{L} = \mathfrak{L}(x_a, y^A, y^A_{a_1}, \ldots, y^A_{a_1 \ldots a_k}).$$

However, it may occur that for some of the y^A fields the order of differentiation is stopped for different values of k. Also, it is possible to separate the set of independent variables into subsets in such a way that some y^A depend only on one of the subsets. These situations may be illustrated by the following possibilities:

1) The set of variables are of type $(x_a, y^A, y^A_{a_1}, \ldots, y^A_{a_1 \ldots a_{s_A}})$ where $1 \leq a \leq n$, $1 \leq A \leq m$, $1 \leq a_1 \leq \ldots \leq a_{s_A} \leq \ldots \leq n$, with $1 \leq s_A \leq k_A$, k_A being the higher order derivative for the y^A.

2) Let i be a partition of the set of integers $\{1, 2, \ldots, n\}$ and j the complementary. Then the set of variables may be of type

$$(x_i, x_j, y^A, y^A_{i_1}, \ldots, y^A_{i_1 \ldots i_{s_A}}, y^A_{j_1}, \ldots, y^A_{j_1 \ldots j_{r_A}})$$

where now $1 \leq s_A \leq k_A$, $1 \leq r_A \leq \ell_A$.

3) We take the above situation where now some partial derivatives of the y's start from an order different from 1 and so the set of variables may be of type:

$$(x_i, x_j, y^A, y^B, y^A_{i_1}, \ldots, y^A_{i_1 \ldots i_{s_A}}, y^A_{j_1}, \ldots, y^A_{j_1 \ldots j_{r_A}},$$
$$y^B_{i_1}, \ldots, y^B_{i_1 \ldots i_{s_B}}, y^B_{j_1}, \ldots, y^B_{j_1 \ldots j_{r_B}}),$$

where $1 \leq s_A \leq k_A$, $1 \leq r_A \leq \ell_A$, $0 < u_B \leq s_B \leq k_B$, $0 < v_B \leq r_B \leq \ell_B$.

The development of a canonical formalism for such situations may be tried. For example, we define the momenta by

$$p_{B+1/a_1 \ldots a_{\beta_B}} = \sum_{j=0}^{k_{B+1} - \beta_B} (-1)^j \frac{d^j}{dx_{a_1} \ldots dx_{a_j}} \left(\frac{\partial \mathfrak{L}}{\partial y^{B+1}_{a_1 \ldots a_j + \beta_B}} \right),$$

where $0 \leq B \leq m-1$, $1 \leq \beta_B \leq k_{B+1}$ $(k_o \overset{\text{def}}{=} 0)$. Let us consider $1 \leq A \leq 2$, $k_1 = 3$, $k_2 = 2$, i.e., $\mathfrak{L}$ is a function of

$$(x_a, y^1, y^2, y^1_{a_1}, y^1_{a_1a_2}, y^1_{a_1a_2a_3}, y^2_{a_1}, y^2_{a_1a_2}).$$

We have $0 \leq B \leq 1$, $1 \leq \beta_o \leq k_1 = 3$, $1 \leq \beta_1 \leq k_2 = 2$.

Therefore:

for $B = 0$

$$p_{1/a_1\ldots a_{\beta_o}} = \sum_{j=0}^{3-\beta_o} (-1) \frac{d^j}{dx_{a_1}\ldots dx_{a_j}} \left(\frac{\partial\Omega}{\partial y^1_{a_1\ldots a_{j+\beta_o}}}\right);$$

for $1 \leq \beta_o \leq 3$, one obtains

$$p_{1/a_1} = \frac{\partial\Omega}{\partial y^1_{a_1}} - \frac{d}{dx_{a_1}}\left(\frac{\partial\Omega}{\partial y^1_{a_1a_2}}\right) + \frac{d^2}{dx_{a_1}dx_{a_2}}\left(\frac{\partial\Omega}{\partial y^1_{a_1a_2a_3}}\right),$$

$$p_{1/a_1a_2} = \frac{\partial\Omega}{\partial y^1_{a_1a_2}} - \frac{d}{dx_{a_1}}\left(\frac{\partial\Omega}{\partial y^1_{a_1a_2a_3}}\right),$$

$$p_{1/a_1a_2a_3} = \frac{\partial\Omega}{\partial y^1_{a_1a_2a_3}}.$$

If $B = 1$

$$p_{2/a_1\ldots a_{\beta_1}} = \sum_{j=0}^{2-\beta_1} (-1)^j \frac{d^j}{dx_{a_1}\ldots dx_{a_j}} \left(\frac{\partial\Omega}{\partial y^2_{a_1\ldots a_j+\beta_1}}\right)$$

and for $1 \leq \beta \leq 2$,

$$p_{2/a_1} = \left(\frac{\partial\Omega}{\partial y^2_{a_1}}\right) - \frac{d}{dx_{a_1}}\left(\frac{\partial\Omega}{\partial y^2_{a_1a_2}}\right)$$

$$p_{2/a_1a_2} = \frac{\partial\Omega}{\partial y^2_{a_1a_2}},$$

(if k_A = constant for all A, we are in presence of the formalism studied before).

4.2. Perturbation theory: an example.

Consider the well known harmonic oscillator equation

$$\ddot{q} + \omega^2 q = 0. \tag{1}$$

The motion of this system can be uniquely determined by giving values to the coordinate and velocity in an arbitrary instant of time t_o. The general solution is obtained from Taylor polynomial expression

$$q(t) = \Sigma \frac{1}{j!} q^{(j)}(t_o)(t-t_o)^j ,$$

where the $q^{(j)}$ for $j > 1$ are obtained from the above equation (1) for given initial values at t_o. Let us take a perturbation of (1) of type

$$\ddot{q} + \omega^2 q = \epsilon \ f(q,\dot{q},t), \tag{2}$$

where ϵ is a small parameter and f is a smooth function standing for some physical effect. The function f has a local character in the sense that f depends directly on the initial values. We may consider a more complex f than in (2). If we include higher derivative in the right-hand side of (2), as $\epsilon \ f(q,\dot{q},t)q^{(j)}$, $1 \le j$ and if we assume a non-local character for the perturbation, $\epsilon \ f(q,\dot{q},t,\bar{t})q^{(j)}$, then we cannot apply the procedure used for the "free equation" (1). In (2) the solutions are, in general, different from the "free solutions" of (1), since for zero values of $f(q,\dot{q},t)$ at a time $\bar{t}$ new singularities appear. We may have infinite values for the higher derivatives $q^{(s)}$, $s \ge j$, when we want to deduce all higher derivatives from the prescribed initial values $q, q^{(1)}, \ldots, q^{(j-1)}$. Even if we look for the solution expanded at a regular point, all higher derivatives $s \geq j$ may contain inverse powers of ε, if the initial values of $q^{(\ell)}$, $0 \leq \ell \leq j-1$, are arbitrarily given and so it may happen that the perturbation theory fails.

A theory in which the higher order derivative of dynamical variables appearing in the equations of motion are included in interaction terms with non-local character was studied in 1955 by T. Taniuti in two extended papers. The author considered also integro-differential equations of type

$$\ddot{q} + \omega^2 q =$$

$$\varepsilon \int_{-\infty}^{\infty} f(t,t_1,\ldots,t_r) g(q(t),\dot{q}(t),q(t_1),\dot{q}(t_1),\ldots,\dot{q}(t_r)) dt_1 \wedge \ldots \wedge dt_r).$$

A complete classical study on local/non-local mechanical theories with linear and non-linear character, as well as field theories, may be found Taniuti's works.

Let us see a simple example given by Taniuti. Consider the Lagrangian

$$L = (1/2) \sum_{i=1}^{2} (\dot{q}_i^2 - \omega_i^2 + q_i^2) + (\varepsilon/2) q_i^{(4)} q_2^2 .$$

The equations of motion are

$$\ddot{q}_1 + \omega_1^2 q_1 = (\varepsilon/2) d^4/dt^4 \, q_2^2$$

$$\ddot{q}_2^2 + \omega_2^2 q_2 = \varepsilon \; q_2 q_2^{(4)} .$$

It results from these equations that

$$\left.\begin{aligned} &q_1^{(j+2)} + \omega_1^2 q_1^{(j)} - (\varepsilon/2) \sum_{r=0}^{j+4} \binom{j+4}{r} q_2^{(r)} q_2^{(n+4-r)} = 0 \\ &q_2^{(j+2)} + \omega_2^2 q_2^{(j)} - \varepsilon \sum_{r=0}^{j} \binom{j}{r} q_2^{(r)} q_1^{(j+4-r)} = 0 \end{aligned}\right\} \quad (*)$$

where $\binom{a}{b} = (a!/b!(a-b)!)$.

PROPOSITION. <u>All higher derivatives</u> $q_i^{(s)}$, $s \geq 2$, $1 \leq i \leq 2$, <u>can be uniquely determined at a given time</u> $t = t_o$ <u>for the prescribed values of</u> $q_i(t_o)$, $\dot{q}_i(t_o)$.

The proof is obtained by writting $q_1^{(j)}$ and $q_2^{(j)}$ in formal power series in ε. Taking

$$q_1^{(j)} = \sum_{\ell=0}^{\infty} a_{j\ell}\, \varepsilon^{\ell}, \qquad q_2^{(j)} = \sum_{\ell=0}^{\infty} b_{j\ell}\, \varepsilon^{\ell}$$

it is possible to show that from the set of equations (*) one has

$$a_{21} = (1/2) \sum_{r=0}^{4} \binom{4}{r} b_{ro}\, b_{4-r\,o}$$

$$b_{21} = b_{oo}\, a_{4o}$$

(it is supposed that $a_{o\ell} = a_{1\ell} = b_{o\ell} = b_{1\ell} = 0$, for $\ell \geq 1$).

All a_{j1}, b_{j1} $(j \geq 2)$ are deduced from the above equations and the substitution of such values into the corresponding equations with $j = 0$, $\ell = 2$ determines $a_{22}, b_{22}, \ldots, a_{j2}, b_{j2}$. Successively applying this procedure we show that all a's and b's are uniquely determined.

4.3. An acceleration dependent Lagrangian for Relativistic Mechanics.

In 1972 F. Riewe proposed an acceleration dependent Lagrangian from which "a relativistic classical theory of spinning particle that follows from a simple Lagrangian and which can be placed in Hamilton's canonical form", would be obtained. The theory of spinning particle had been developed in 1926 by Frenkel and Thomas (see Riewe (1972)) but in this approach there was no Hamiltonian form for the theory. The inclusion of an acceleration term in the Lagrangian, as we have seen in Bopp-Podolsky's electromagnetism, allows for a more simple quantization.

Riewes' Lagrangian is

$$L = -(1/2)m\ c^2\ (\dot{x}^{\alpha}\dot{x}_{\alpha} - \ddot{x}^{\alpha}\ddot{x}_{\alpha}/\omega^2) \tag{3}$$

where m and ω are taken as constants. In his original paper, the following relativistic notation is used: $x^1 = x$, $x^2 = y$, $x^3 = z$, $x^4 = ict$, with $x_i = x^i$ and $x_4 = -x^4$, the proper time s is given by

$$c^2(ds)^2 = -dx^u dx_u = -(dx^1)^2 - (dx^2)^2 - (dx_3)^2 + (dx^4)^2.$$

REMARK (1). We recall that the usual (covariant) Lagrangian formalism for a relativistic non-spinning particle is obtained from the Lagrangian

$$(1/2)m\ \dot{x}^u\dot{x}_u\ .$$

The quantization of the corresponding Hamiltonian leads to the Klein-Gordon equation, which is of second order. We will see that in the present case the wave equation is of first order in time.

From Lagrangian (3) we obtain the corresponding equation of motion

$$(d^2/ds^2)\ddot{x}^u + \omega^2\ddot{x}^u = 0, \tag{4}$$

with general solution given by

$$x^u = x_o^u + v_o^u s + \ell_o^u \sin(ws + \theta_o^u), \tag{5}$$

where ℓ and θ are values to be taken as initial conditions. This solution suggests that (4) describes a particle in simple harmonic motion about a moving point. From this fact and considering a particular Lorentz frame and a choice of initial conditions, Riewe shows that the spatial and time components of (5) become

$$\vec{r} = \vec{r}_o + \vec{v}_o s + \ell_o \; (i \cos ws + j \sin cos) \tag{6}$$

$$t = t_o + v_o^4 s.$$

For a sufficiently small value of ℓ_o (6) describes a spinning particle with an interval angular momentum in the z-direction and so "the particle is automatically endowed with a spin due to its orbital motion withoug the need to postulate an additional intrinsic spin" (Riewe (1972)).

The Hamiltonian formalism is obtained by the Ostrogradsky procedure as it was shown for the Generalized Electromagnetism. It is found that the values of the momenta are

$$p_{1/u} = m \; v_{ou} \; ; \qquad p_{2/u} = \ell_{ou} \sin(ws + \theta_o^{(u)})$$

with Hamiltonian

$$H = q_2^u \, p_{1/u} - (1/2)m \; q_2^u \, q_{2u} - \omega^2 \; p_{2/u} \; p^{2/u}/2m,$$

where the coordinates are defined by:

$$q_1^u = x^u, \quad q_2^u = \dot{x}^u, \quad q_{2u} = \dot{x}_u ,$$

$$p_{2/u} = \partial L/\partial \ddot{x}^u, \quad p^{2/u} = \partial L/\partial \ddot{x}_u , \qquad \text{etc}$$

From the general solution, the values of the momenta shown that $p_{1/u}$ has the same form as the usual ones and $p_{2/u}$ gives the instantaneous position of the particle relative to its center of mass.

Riewe's approach can be applied in a more general situation with the help of Dirac-Bergmann's constraint theory. This whas shown by J.R. Ellis in 1975, inspired in a paper of S. Shanmugadhasan (1973) on degenerate Lagrangians.

Ellis' procedure consists in transforming the original order 2 Lagrangian into a new one containing extra coordinates and in such a way that the acceleration "desapears" (in the sense that it is taken as a new variable). It follows that the augmented Lagrangian has no velocities corresponding to this set of coordinates. So the Hessian matrix contains a column of zeros and the new Lagrangian is naturally degenerate in the sense of Dirac constraint theory.

Ellis' study for relativistic Lagrangians is also subjected to the requirement that $\dot{x}^u \dot{x}_u = 1$. The augmented Lagrangian is obtained from (4) by inserting the variables λ_1, λ_2^u, λ_3^u, $\dot{\lambda}_2^u$ so that

$$\bar{L} = L(x^u, \dot{x}^u, \dot{\lambda}_2^u) + \lambda_3^u(\dot{x}_u - \lambda_2 u) + (1/2)\lambda_1(\dot{x}^u \dot{x}_u - 1)$$

$$= (1/2)(\lambda_1 - A)\dot{x}^u \dot{x}_u + (1/2)B\, \dot{\lambda}_2^u \dot{\lambda}_{2u} + \lambda_3^u(\dot{x}_u - \lambda_{2u}) - (1/2)\lambda_1 ,$$

where the $\ddot{x}^u$ of the original Lagrangian are replaced by $\dot{\lambda}_2^u$ and the multipliers λ_1 and λ_3^u are considered for the constraint relations $\dot{x}^u \dot{x}_u = 1$ and $\lambda_2^u = \dot{x}^u$, $(\lambda_{2u} = \dot{x}_u)$. The values A, B are defined by

$$A = mc^2, \qquad B = mc^2/\omega^2.$$

A straightforward calculation shows that the Euler-Lagrange equations differ from (4) by terms of type

$$\dot{x}^\nu \ddddot{x}_\nu \dot{x}^u + (\lambda_1/B)\ddot{x}^u$$

due to the requirement that the Lagrangian must be consistent with the condition $\dot{x}^u \dot{x}_u = 1$.

From the definition of the augmented Lagrangian we see that the Hessian matrix is of rank eight and the Lagrangian is degenerated. To develop the Hamiltonian formalism we define the momenta conjugate to the coordinates by

$$p^u = -(\partial\bar{L}/\partial\dot{x}_u), \qquad p_1 = -(\partial\bar{L}/\partial\lambda_1)$$
$$p_2^u = -(\partial\bar{L}/\partial\lambda_{2u}), \quad p_3^u = -(\partial\bar{L}/\partial\lambda_{3u})$$

and from the fact that the corank of the Hessian is five we define the following first kind subsidary conditions (in the sense of Dirac):

$$f_{(1)} = p_1 \approx 0, \qquad f_{(2)}^u = p_3^u \approx 0,$$

where $\approx$ is the symbol used by Dirac to denote that they are "weak" equations.

In the rest of his paper Ellis examines the conditions for the introduction of further constraint relations to define the Hamiltonian and concludes that "we have been unable to compare our Hamiltonian with Riewe as they are so widely different, but it is sufficient to say that the Hamiltonian that we have constructed satisfies Hamilton equations in all the multipliers as well as the original variables x^u, p^u, λ_2^u, p_2^u (for an acceleration dependent Lagrangian), so that it is consistent with the condition $\dot{x}^u\dot{x}_u = 1$, whereas Riewe's Hamiltonian is constructed from the original variables x^u, p^u, $\dot{x}^u$, $\partial L/\partial \ddot{x}_u$ only, and is, therefore, not consistent with the condition $\dot{x}^u\dot{x}_u = 1$" (Ellis (1975)).

4.4. Quantization of higher order field theories.

As we have seen before, the study of the canonical form of the Euler-Lagrange equations for higher order Lagrangians can be approached in different manners. Due to the non-existence of a unique method to obtain the Hamiltonian formalism, the quantization of theories involving higher order derivatives remains an open question. Some authors have proposed quantization schemes based on Jacobi-Ostrogradsky's Hamiltonization method. Thus, the canonical quantization of Generalized Particle Mechanics via Jacobi-Ostrogradisky could be characterized by a wave function depending on the variables $q_a^{(j)}$ and the momenta could be replaced by the operator

$$\frac{\hbar}{i}\left(\frac{\partial}{\partial q_a^{(j)}}\right).$$

Unlike in the usual theory, the possibility of choosing among all variables those which will play the role of canonical coordinates does not allow an a priori determination of which variables are to be replaced by operators. Although the choice of an appropriate Lagrangian does this determination, some divergent interpretations will remain.

In what follows, we will present, as an illustration, some attempts of quantization.

We have seen before (§3.5) that Hayes and Jankowski (1968) and Hayes (1969) worked in the direction of showing the existence of ambiguity in the generalized formalism (non unicity of the Lagrangian) but Kimura (1972) placed the formalism in the constraint theory, eliminating the objections Hayes and Jamkowski also presented an attempt of quantization for non-degenerate Lagrangians of higher order; Kimura worked out the degenerate case. All these papers concern Particle Mechanics.

The Poisson bracket for Generalized Mechanics is

$$\{u,v\} = \left(\frac{\partial u}{\partial q_a^{(j)}} \frac{\partial v}{\partial p_{a/j+1}} - \frac{\partial u}{\partial p_{a/j+1}} \frac{\partial v}{\partial q_a^{(j)}}\right)$$

(and we find, by analogy with the usual procedure, that

$$\{q_a^{(s)}, q_s^{(j)}\} = \{p_{a/s+1}, p_{b/j}\} = 0; \quad \{q_b^{(s)}, p_{a/s+1}\} = \delta_{ba}\,\delta_{sj+1}$$

$$\{q_a^{(j)}, H\} = \partial H/\partial p_{a/j+1}; \quad \{p_{a/j+1}, H\} = -\partial H/\partial q_a^{(j)}.$$

Comparing these two last equations with the canonical equations we are led to postulate that

$$\{p_{a/j+1}, q_b^{(s)}\} = p_{a/j+1} q_b^{(s)} - q_b^{(s)} p_{a/j+1}$$

$$\stackrel{\text{def}}{=} \hbar/i\; C(j+1,k)\;\delta_{j+1}\,\delta_{s+1}\,\delta_{ab}$$

where $C(j+1,k)$ is a c-number. Let us take F to be any

term of the Hamiltonian of the form

$$F = \prod_{a=1}^{n} \prod_{j=0}^{k} (p_{a/j+1})^{\alpha(a,j+1)} (q_a^{j+1})^{\beta(a,j+1)} .$$

Then we find

$$\{q_a^{(j)}, F\} = -\hbar/i \; C(j+1,k)(\partial F/\partial p_{a/j+1})$$

$$\{p_{a/j+1}, F\} = \hbar/i \; C(j+1,k)(\partial F/\partial q_a^{(j)})$$

and

$$\{q_a^{(j)}, H\} = -\hbar/i \; C(j+1,k)(\partial H/\partial p_{a/j+1})$$

$$\{p_{a/j+1}, H\} = \hbar/i \; C(j+1,k)(\partial H/\partial q_a^{(j)})$$

Hayes says that these two last equations may be considered a the generalized Heisenberg equations of motion for Bose-Einstein quantization. If we take $j = 0$ and $k = 1$ we reduce to the standard situation. It is not hard to see that the commutator of $p_{a/j+1}$ and $q_{a/j}$ commutes with H. The validity of the above postulate rests upon the consequent experimental implications and for this reason Hayes has not taken $C(j+1,k) = 1$. Also, if $\{p_{a,j+1}, q_b^{(s)}\}$ is acting on a wave function, we can take

$$p_{a/j+1} \to \hbar/i \; C(j+1,k)(\partial/\partial q_a^{(j)}) \quad \text{and} \quad q_k^{(j)} \to q_k^{(j)}$$

or

$$p_{a/j+1} \to p_{a/j+1} \quad \text{and} \quad q_a^{(j)} \to \hbar/i \; C(j+1,k)\partial/\partial p_{a/j+1} .$$

It is also possible to see (see our discussion on Hayes paper in §3.5) that there is a change in the relation of the generalized momenta and the corresponding operator. It is possible to show that the configuration variable is canonically conjugated to $2p$ and p must be replaced by $(1/2)(\hbar/i)(\partial/\partial q)$ and $C(1/2)$ is 1/2. Therefore "there is a change in the definitions of the momenta which results is a different form for the uncertaint principle and the rules for quantization" (Hayes (1969)).

4.5. Bornea's quantization for linear momenta.

Adopting the same principles of standard quantum theory, M. Borneas obtained in 1972 a quantum equation of motion with higher derivatives. However his method is applicable when all momenta are linear with respect to the set $\{q_A^{(i)}\}$ of canonical Ostrogradsky variables. For example, if $L = (m/2)[q^{(2)}]^2$ then Ostrogradsky momenta are $p_o = -m\, q^{(3)}$ and $p_2 = m\, q^{(2)}$.

Let us consider a Lagrangian of second order. Then $p_{A/1}$ will be a function of the q_A's and their derivatives with respect to the time up to third order and $p_{A/2}$ up to second order. In general $q_A^{(3)}$ cannot be eliminated (in Hayes' procedure, the particular form of the Lagrangian makes it possible, with the help of the equation of motion). To obtain a quantization, Borneas introduced "reduced momenta" to replace Jacobi-Ostrogradsky's momenta. So, assuming the linear property and deriving the energy function $E = p_{A/1}\, q_A^{(1)} + p_{A/2}\, q_A^{(2)} - L$ with respect to $q_A^{(1)}$, $q_A^{(2)}$ we have

$$\frac{\partial^2 E}{\partial q_A^{(1)2}} = \frac{\partial p_{A/1}}{\partial q_A^{(1)}} - \frac{\partial}{\partial q_A^{(1)}}\left[\frac{d}{dt}\left(\frac{\partial L}{\partial q_A^{(2)}}\right)\right]$$

$$\frac{\partial^2 E}{\partial q_A^{(2)^2}} = -\frac{\partial p_{A/2}}{\partial q_A^{(2)}} + \frac{\partial}{\partial q_A^{(2)}} \left(\frac{\partial L}{\partial q_A^{(2)}}\right)$$

It is clear now that the momenta may be written in the form

$$p_{A/j} = \int_0^{q_A^{(j)}} -\frac{\partial^2 E}{\partial q_A^{(j)2}} \, dq_A^{(2)} + \dots$$

and we define the "reduced momenta" $\hat{p}_{A/j}$ as being the first term in the right hand side of the above equality, i.e., $\hat{p}_{A/j}$ represent a part of the Ostrogradsky's canonical conjugate momenta. We express the energy by the reduced momenta substituting the derivatives up to order two and E is prepared for quantization. For this, we introduce the "reduced momentum operator"

$$\hat{p}_{A/j} = -i \, \hbar \, a_{A/i2} \, \partial/\partial q_A^{(j-1)} + b_{A/i2} \,,$$

where a and b are operators to be determined under certain considerations.

Let us consider an example: take $m \dot{q}^2/2$ for the Lagrangian and consider an extension of second order of type

$$L = c_o \dot{q}^2 + c_2 \ddot{q}^2 .$$

The Ostrogradsky momenta are

$$p_1 = 2c_1 \dot{q} = 2c_2 \dddot{q} \,; \qquad p_2 = 2c_2 \ddot{q}$$

and the "reduced momenta" are

$$\hat{p}_1 = 2c_1 \dot{q} \,; \qquad \hat{p}_2 = 2c_2 \ddot{q} \,.$$

The introduction of the "reduced operators" leads to

$$\hat{E} = -\frac{\hbar^2}{4c_1} a_{12} \frac{\partial^2}{\partial q^2} - \frac{\hbar^2}{4c_2} \frac{a_{22}^2 \partial^2}{\partial \dot{q}^2} + (4c_1)^{-1} \{-i \hbar A_{12} (\frac{\partial}{\partial q}), b_{12}\}_+$$

$$+ (4c_2)^{-1} \{-i \hbar a_{22} (\frac{\partial}{\partial \dot{q}}), b_{22}\}_+ - \frac{c_2 \ddot{q}}{c_1} \frac{-i \hbar a_{12} \partial}{\partial q} +$$

$$+ b_{12} + \frac{b_{12}^2}{4c_1} + \frac{b_{22}^2}{4c_2},$$

where $\{ , \}_+$ is the anticommutator of the enclosed terms. Assuming now that the eigenvalues of $\hat{E}$ are real ($\hat{E}$ is Hermitian), that $\hat{E}$ commutes with the operators and that the wave function does not contain $\dddot{q}$ we determine the conditions to be fulfilled by the a's and b's. The simplest solution is all terms except the first two to be null in the above expression, with the supplementary condition

$$a_{12}^2 = a_{22}^2 = 1.$$

We have the following equation:

$$-(\hbar^2/4c_1)(\partial^2\psi/\partial q^2) - (\hbar^2/4c_2)(\partial^2\psi/\partial \dot{q}^2) = E\psi.$$

Puutting $\hat{E} = i \hbar \partial/\partial t$ and adding the potential energy operator U we have the quantum equation of motion:

$$-(\hbar^2/4c_1)(\partial^2\psi/\partial q^2) - (\hbar^2/4c_2)(\partial^2\psi/\partial \dot{q}^2) + U\psi = i \hbar \partial\psi/\partial t.$$

The extension to higher dimensions is simple. Let us now examine solutions for the harmonic oscillator equation (Borneas (1973)). We shall put $c_1 = m/2$, $c_2 = m/2 \omega^2$ and $U = m \omega^2 q^2/2$ and we will seek solutions of the form

$$\psi = \frac{\partial\psi}{\partial t} \frac{\partial^2\psi}{\partial q \partial \dot{q}} = \psi_t \psi_{q\dot{q}},$$

where ψ_t depends uniquely on t, $\psi_{q\dot{q}}$ on q and $\dot{q}$. We write

$$\frac{1}{\psi_{q\dot{q}}}\left(-\frac{\hbar^2}{2m}\frac{\partial^2\psi_{q\dot{q}}}{\partial q^2}-\frac{\omega^2\hbar^2}{2m}\frac{\partial^2\psi_{q\dot{q}}}{\partial\dot{q}^2}\right)=\frac{i\hbar}{\psi_t}\frac{\partial\psi_t}{\partial t}=E,$$

which gives

$$\psi_t = \text{const.}\ \exp\left(-\frac{i}{\hbar}Et\right).$$

If we look for solutions of the form

$$\psi_{q\dot{q}} = \psi_q \cdot \psi_{\dot{q}},$$

we obtain the equations

$$-\frac{\hbar^2}{2m}\frac{d^2\psi_q}{dq^2}+\frac{m\,\omega^2 q^2}{2}\psi_q - E\psi_q = \bar{E}\psi_q \qquad (7)_1$$

$$\frac{-\omega^2\hbar^2}{2m}\frac{d^2\psi_{\dot{q}}}{d\dot{q}^2} = E\psi_{\dot{q}} \qquad (7)_2$$

and therefore

$$E + \bar{E} = \hbar\,\omega\left(n+\frac{1}{2}\right), \qquad n = 0,1,2,\ldots \qquad (8)$$

$$\psi_q = \text{const}\ H_n\left(x\left(\frac{m\omega}{\hbar}\right)^{1/2}\right)\exp\left(-\frac{m\omega}{2\hbar}q^2\right),$$

where H_n are the Hermite polynomials. Equation $(7)_2$ has solutions of type

$$\psi_{\dot{q}} = \text{const.}\ \exp\left(\pm\frac{i}{\hbar}\left(\frac{2m\bar{E}}{\omega^2}\right)^{1/2}\dot{q}\right). \qquad (9)$$

From (8) and (9) and the condition $\psi_{q\dot{q}} = \psi_q\,\psi_{\dot{q}}$ one obtains the solution of our equation:

$$\psi_{q\dot{q}} = N\exp\left(\pm\frac{i}{\hbar}\left(\frac{2m\bar{E}}{\omega^2}\right)^{1/2}\dot{q}\ H_n\right)\exp\left(-\frac{m\omega}{2\hbar}q^2\right),$$

where N is a normalization factor. The energy of the oscillator is

$$E = \hbar\,\omega(n+1/2) - \bar{E}.$$

4.6. Difficulties with the higher order theories.

As we have pointed out before Pais and Uhlenbeck (1950) showed that theories with higher derivatives might have better convergence properties than the standard theory. However they say that "it is very difficult (...) to reconcile the three requirements of convergence, positive definiteness and strict causality..." (for Physical problems).

The study of those authors starts theoretically with an equation of type

$$F(\circledcirc)g = h,$$

where F is taken as a polynomial function of a certain degree in the operator $\circledcirc$ and g, h are functions. For example, if we consider a (one dimensional) Mechanical system then $\circledcirc$ is taken to the (d/dt), F a polynomial with degree equal to the highest order of derivation $g = q$ is the coordinate function and $h \equiv 0$. For a Field theory we take $\circledcirc$ as the D'Alembertian operator $\Box$, $g = \Box A_u$, $h = -j_u$ (the current vector) and $F(\Box) = \prod_i (1-\Box/u_i^2)$.

Let us consider the Mechanical model. Pais and Uhlenbeck chose an approach which is different from Jacobi-Ostrogradsky's Method for quantization of the theory. They remarked that Jacobi-Ostrogradsky's method cannot be applied when F is a formal power series. Also it is impossible to separate the Jacobi-Ostrogradsky Hamiltonian in a product of some linear combination of harmonic oscillator Hamiltonians, which would make the quantization simpler. They proposed to obtain this separation by introducing appropriate coordinates, called "oscillator coordinates" defined by

$$Q_i = \prod_j (1+D^2/\omega_j^2)\, q \qquad (D = d/dt)$$

where the ith factor is missing in the product, q is the original coordinate, ω_j the frequencies of oscillations. The polynomial F is defined by

$$F = \prod_i (1+D^2/\omega_i^2).$$

With this method, it is possible to show that the Hamiltonian is some linear combination of oscillator Hamiltonians, some with positive-definite energy and some with negative-definite energy. The analysis of all possibilities, that is, real or complex, single or multiple frequencies ω_i, leads to the conclusion that "if the Lagrangian L contains more than one frequency the energy is always indefinite, both classically and quantum mechanically (...). It is this indefinitess which will constitute a grave difficulty in field theoretical applications."

After these considerations the authors examine the equations of motion of infinite order, that is, F contains derivatives of infinite order. They are able of putting the problem in a form which corresponds to a system with only one frequency and exponential factor. They apply Heisenberg's method to this system.

Some aspects of Field theory are also developed in the text and they arrive to show that the self-energy of the electron is finite. But this result is obtained by sacrifying the possibility of creating quanta of negative energy, leading to an inconsistent theory where no physical definition of the vacuum is possible.

III.5 - Exercises

1. Let G_k be generated by the forms $\theta^A, \theta^A_{a_1}, \ldots, \theta^A_{a(k-1)}$, where the θ's are the components of the canonical form θ_k on J^kM; G is called the <u>contact system</u> on J^kM. The <u>contact</u> n-<u>form system</u> of J^kM is $G_{k,n} = \{G_k \wedge \omega_a\}$, $\omega_a = (-1)^{(a-1)} dx_1 \wedge \ldots \wedge \widehat{dx}_a \wedge \ldots \wedge dx_n$. $Ver^{(r)}(J^kM)$ is the set of all tangent vectors of J^kM which are vertical over J^rM, $r < k$. Define S_k and $\hat{G}_k$ by

$$S_k = \{v \in G_{k,n};\ i_X\, dv \equiv 0 \mod G_{k,n},\ \forall X \in Ver^{(k-1)}(J^kM)\}$$

and

$$\hat{G}_o = G_{1,n}, \quad \hat{G}_k = \hat{G}_{k-1} + G_{k,n}/S_k.$$

   (i) Show that $\hat{G}_k$ is generated by $\{\theta^A_{a(i-1)} \wedge \omega_{a_i}\}$, $1 \le i \le k-1$ and that if $u \in Sec(J^kM)$ then $u^*\hat{G}_k = 0$ if $u = j^k(\beta^k \circ u)$.

   (ii) If Ω is a n-form on $J^{2k-1}M$ such that $\Omega = \mathfrak{L}\omega \mod \hat{G}_{2k-1}$ and $i_X\, d\Omega \equiv 0 \mod G_{2k-1,n}$ for every vertical vector field X on $J^{2k-1}M$ along N then Ω is unique (where $\mathfrak{L}: J^kM \to \mathbb{R}$ is smooth).

   (iii) Let $\mathfrak{L}$ be a Lagrangian and suppose that the characteristic matrix of $\mathfrak{L}$ is of maximal rank. Show that if $u \in Sec(J^{2k-1}M)$ is such that for every $X \in Ver(J^{2k-1}M)$, $u^*\, i_X\, d\Omega = 0$ then the section $s = \beta^{2k-1} \circ u$ is an extremal for $\mathfrak{L}$ and $\rho^{2k-1}_k \circ u = \tilde{s}^k$.

(iv) Let f be a section of $(J^{2k-1}M, \rho^{2k-1}_{k-1}, J^{k-1}M)$ and $s \in Sec(M)$. Suppose that $(\tilde{s}^{k-1})^*(i(X)f^*\Omega) = 0$ for every vertical vector field X on $J^{k-1}M$. If $u = f \circ \tilde{s}^{k-1}$ is such that $\rho^{2k-1}_{k} \circ u = \tilde{s}^{k}$ then show that $u^*(i_Y \Omega) = 0$ for every vertical vector field Y on $J^{2k-1}M$ along N and s is an extremal. If $f^*d\Omega = 0$ then at least locally $f^*\Omega = dF$. Show that this equality leads to the generalized Hamilton-Jacobi equations

$$\frac{\partial F^a}{\partial x_a} + H(x_a, y^A, \ldots, z^A_{a(k-1)}, \frac{\partial F^a}{\partial y^A}, \ldots, \frac{\partial F^a}{\partial z^A_{a(k-1)}}) = 0$$

where $F = F^a \omega_a$. (For more details for (i)-(iv) see Shadwick (1982)).

2. Let (M,p,N) be a fibered manifold. Then J^kM is an affine bundle over $J^{k-1}M$ whose associated vector bundle is $S^kT^*N \underset{J^kM}{\otimes} Ver(M)$ (which can be identified with $Ver(J^{k-1}M)$), where S^k is the symmetric tensor product. We recall that if (A,π,B) is a vector bundle over B and (C,p,B) is a fibered manifold such that (D,f,C) is a vector bundle over C then the tensor product of D and the reciprocal image of A over C is denoted by $A \otimes D$ (for details see Pommaret (1982), p.34).

Definition (Muñoz (1984)). A vector field X on J^kM is said to be an infinitesimal contact transformation (i.c.t.) of order k if for every connection ∇ on $Ver(J^{k-1}M)$ there is an endomorphism f of $Ver(J^{k-1}M)$ such that

$$\mathcal{L}_X \theta_k = f \circ \theta_k ,$$

where $\mathcal{L}$ is the Lie derivative with respect to the connection ∇ (see Choquet-Bruhat (1968)) and θ_k is the canonical form of order k. Show that

(i) The above definition does not depend on the choice of the connection.

(ii) For every vector field X on M there is a unique $\bar{X}^k$ i.c.t. projectable on X. Show also that all i.c.t. form a Lie subalgebra of vector fields on J^kM.

(iii) Suppose that a vector field X on M is projectable along N with projection $\bar{X}$. Let X_t, $\bar{X}_t$ be the corresponding flows, (in general only locally defined). Then

$$X_t^{(k)}(\tilde{s}^k(x)) = j^k_{\bar{X}_t(x)}(X_t \circ s \circ \bar{X}_{-t})$$

is the flow generated by $\bar{X}^k$.

(iv) If Y is a i.c.t. of J^kM (projectable along N) then there is a vector field X on M (projectable along N) such that $\bar{X}^k = Y$, (for more details, see Garcia & Muñoz (1983), Muñoz (1984)).

3. Let (M,p,N) be a trivial fibered manifold. The set Sec(M) admits a structure of $C^\infty(N)$-modulus and therefore Sec(M) is locally a free modulus of rank m (dim M = n+m), since the configuration bundle (M,p,N) is supposed to be trivial. Thus, for every point $x \in N$ there is an open set U of N, containing x and m sections s^A, $1 \leq A \leq m$ such that for each V in U, $x \in V$, the sections s^A along V form a basis for $\mathrm{Sec}_V(M)$ over the algebra $C^\infty(V)$. Defining

$$C^{\infty}_{\alpha^1}(J^1M,M) = \{f \in C^{\infty}(J^1M,M);\ p\circ f = \alpha^1\}.$$

We see that this set has a structure of $C^{\infty}(J^1M)$-modulus over the algebra $C^{\infty}(J^1M)$ and $\{s^A\}$ induces a basis for $C^{\infty}_{\alpha^1}(H^1M,M)$ along $(\alpha^1)^{-1}(V)$ defined by $F_A = s^A\circ\alpha^1$.

(i) Verify if it is possible to extend such procedure for higher order jets (for example, take (J^1M,α^1,N) as being now the configuration bundle).

(ii) In that point of view is there a way to construct locally the canonical form θ_k ? (see P.M. Garcia (1968), Rodrigues (1977)).

4. Show the local expression $(1)_1$ for the formal derivative given in Definition (1) of Chapter III.

5. Derive the Hamilton-Jacobi method for the situations described in §2.4 and §2.6 of Chapter III (see Musicki (1978)) and Noether's theorem for Generalized Mechanics (see Anderson (1973)). Try also to develop the formulation on integral invariants of higher order (Musicki (1978)).

6. Consider the operator $D^i = \partial^i/\partial x^i$ and put D^{-i} for the inverse of D^i. Try to develop a theory which includes negative derivatives in the Lagrangian: $L = L(x,D^{-k}y,\ldots,y,$ $,\ldots,D^k y)$. This situation occurs when we deal with integro-differential equations (see Anderson (1973)(b), Ince (1944), Ordinary Differential Equations, N.Y., p. 138).

7. Verify if it is possible to develop a formalism for Lagrangians like in §4.1 of Chapter III in terms of Jet manifolds.

APPENDIX A

VECTOR BUNDLES

A fibered manifold (E,p,B) is called a <u>vector bundle over</u> B with <u>fiber</u> F if there exist an open covering $\{U_i\}_{i\in I}$ of B and diffeomorphisms $h_i\colon E_{U_i} = \bar{p}^1(U_i) \to U_i \times F$ such that the following diagram:

$$\begin{array}{ccc} E_{U_i} = \bar{p}^1(U_i) & \xrightarrow{\;h_i\;} & U_i \times F \\ & {\scriptstyle p}\searrow \quad \swarrow{\scriptstyle p_1} & \\ & U_i & \end{array}$$

is commutative, p_1 being the canonical projection, and the restriction $h_i/E_b = \bar{p}^1(b)$ of h_i to any fiber E_b is a linear isomorphism, $b \in U_i$.

$E_b = \bar{p}^1(b)$ is called the <u>fiber</u> of E <u>over</u> b and the couple (U_i, h_i) a local <u>trivialization</u> of E over U_i.

Let (U_i, h_i) and (U_j, h_j) be two local trivialization of E such that $U_i \cap U_j \neq \emptyset$. Then the mapping

$$h_j \circ h_i^{-1}\colon (U_i \cap U_j) \times F \to (U_i \cap U_j) \times F$$

have the following form

$$(b,v) \rightsquigarrow (b, h_{ji}(b)v),$$

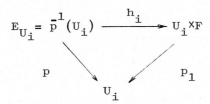

when $h_{ji}(b)$ is an automorphism of F. It follows that the mapping $h_{ji}\colon U_i \cap U_j \to GL(F)$ is differentiable, $GL(F)$ being the Lie group of the automorphisms of F; the mapping h_{ji} will be called the <u>transition function</u> of E on $U_i \cap U_j$. We can easily prove that

$$\left.\begin{array}{l} h_{ii} = id_F, \quad \text{on } U_i \\ h_{kj} \circ h_{ji} = h_{ki}, \quad \text{on } U_i \cap U_j \cap U_k \end{array}\right\} \qquad (1)$$

Now, suppose that there exist an open covering $\{U_i\}_{i \in I}$ of a manifold B and a real vector space F such that there is a differentiable mapping $h_{ji}: U_i \cap U_j \to GL(F)$ for any couple of U_i and U_j with non-empty intersection. If the mappings h_{ji} satisfy (1) then we can define a vector bundle (E,p,B) with fiber F such that its transition functions are precisely h_{ji}. Indeed, E is the quotient set of the disjoint union $\bigcup_{i \in I} (U_i \times F)$ under the equivalence relation according to the pair $(b,v) \in U_i \times F$ is related to the pair $(b',v') \in$ $\in U_j \times F$ if and only if $b' = b$ and $v' = h_{ji}(b)v$, and the projection p is the obvious one.

Now, let (E,p,B) and (E',p',B') be vector bundles. By a vector bundles homomorphism between (E,p,B) and (E',p',B') we mean a couple (f,f_o) of maps $f: E \to E'$ and $f_o: B \to B'$ such that the following diagram

$$\begin{array}{ccc} E & \xrightarrow{\ f\ } & E' \\ {\scriptstyle p}\downarrow & & \downarrow{\scriptstyle p'} \\ B & \xrightarrow{\ f_o\ } & B' \end{array}$$

is commutative and the restriction f_b of f to any fiber E_b is a homomorphism of vector spaces.

We shall simply say that $f: E \to E'$ is a <u>homomorphism</u>

of vector bundles <u>over</u> f_o.

In particular, if (E,p,B) and (E',p',B') are vector bundles over the same manifold B and $f_o = id_B$, then $f: E \to E'$ will be called a <u>homomorphism over</u> B.

Let (U,h) and (U',h') be local trivializations of E and E', respectively, such that $f_o(U) \subset U'$. Then the mapping

$$h' \circ f \circ h^{-1}: U \times F_1 \to U' \times F_2,$$

(F_1 and F_2 being the fiber of E and E', respectively) have the form

$$(b,v) \rightsquigarrow (f_o(b), \tau(b)v),$$

where $\tau(b)$ is a linear mapping from F_1 into F_2. It is easy to check that the mapping

$$U \to \mathfrak{L}(F_1,F_2)$$
$$b \rightsquigarrow \tau(b),$$

is differentiable, where $\mathfrak{L}(F_1,F_2)$ is the vector space of linear mappings from F_1 into F_2.

We have the following:

PROPOSITION (1). <u>Let</u> $f: E \to E'$ <u>and</u> $g: E' \to E''$ <u>be vector bundle homomorphisms</u>. <u>Then</u> $g \circ f: E \to E''$ <u>is a vector bundle homomorphism</u> (<u>over</u> $g_o \circ f_o$, <u>if</u> f <u>and</u> g <u>are homomorphism over</u> f_o <u>and</u> g_o, <u>respectively</u>).

Then a vector bundle homomorphism $f: E \to E'$ over f_o will be called an <u>isomorphism</u> if there exists a vector bundle homomorphism $g: E' \to E$ over g_o such that $g \circ f = id_E$ and $f \circ g = id_{E'}$ (and, therefore, $g_o \circ f_o = id_B$, $f_o \circ g_o = id_{B'}$, if

E and E' are vector bundles over B and B', respectively). In such a case, we will say that E and E' are <u>equivalent</u>. Obviously, $f\colon E \to E'$ is a vector bundle isomorphism if and only if the restriction f_b of f to any fiber E_b is a linear isomorphism.

EXAMPLE. Let $f\colon B \to B'$ be a differentiable mapping and $Tf\colon TB \to TB'$ the induced mapping between the tangent bundles of B on B'. Then Tf is a vector bundle homomorphism. Moreover, Tf is an isomorphism if and only if f is a diffeomorphism.

Now, let (E,p,B) be a vector bundle with fiber F and $f\colon B' \to B$ a differentiable mapping. Let $E' = B' \times_B E'$ the set of the ordered pairs (b',e) such that $f(b') = p(e)$. Denote by p' and f' the restrictions to E' of the canonical projections of $B' \times E$ over B' and E, respectively. Then we have the following commutative diagram:

$$\begin{array}{ccc} E' & \xrightarrow{\;f'\;} & E \\ {\scriptstyle p'}\downarrow & & \downarrow{\scriptstyle p} \\ B' & \xrightarrow{\;f\;} & B \end{array}$$

and we can easily check that (E',p',B') is a vector bundle over B' with fiber F and $f'\colon E' \to E$ is a vector bundle homomorphism over f. E' will be called the <u>induced bundle</u> of (E,p,B) via f.

We have the following universal property

PROPOSITION (2). <u>Let</u> (E'',p'',B') <u>be a vector bundle and</u> (f'',f) <u>a vector bundle homomorphism between</u> E'' <u>and</u> E' <u>over</u> f. <u>Then there exist a unique vector bundle homomorphism</u> $\tilde{f}$

from E'' *into* E' *over* B' *such that the following diagram is commutative:*

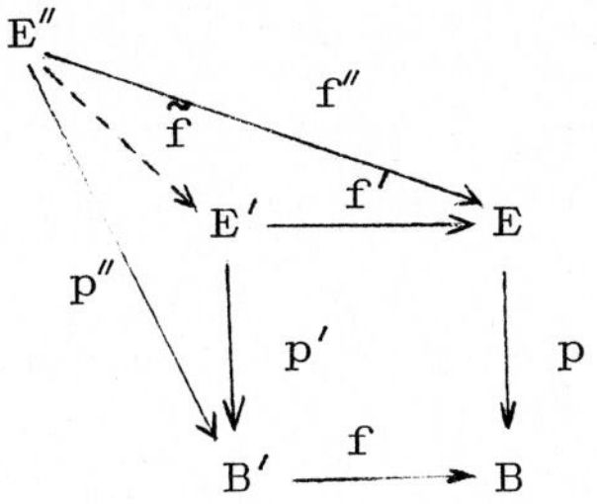

Proof: Indeed, $\tilde{f}$ is defined by

$$\tilde{f}(e'') = (p''(e''), f''(e'')). \qquad \square$$

Now, let E and E' be a vector bundles over B and $f\colon E' \to E$ a vector bundle homomorphism such that the restriction $f_b\colon E_b \to E'_b$ of f to any fiber E_b is injective. We say that E' is a *vector subbundle* of E.

In such a case, we can identify the fiber F of E to $F' \times F''$, where F' is the fiber of E' and F'' is some real vector space. Moreover, we can choose an open covering $\{U_i\}_{i\in I}$ of B with local trivializations h_i and h'_i over U_i of E and E', respectively, such that

$$h_i \circ f \circ h_i'^{-1}\colon U_i \times F' \to U_i \times F' \times F''$$

have the form

$$(b,v) \rightsquigarrow (b,v,0).$$

Obviously, we can identify E' to $f(E') \subset E$ if there can be no confusion.

Let now (E',p',B') and (E,p,B) be a vector bundles and $f\colon E' \to E$ a vector bundle homomorphism over B. Then

$$\operatorname{Ker} f = \bigcup_{b \in B} \operatorname{Ker} f_b$$

is a vector subbundle of E' which will be called the <u>kernel</u> of f and

$$\operatorname{Im} f = \bigcup_{b \in B} \operatorname{Im} f_b$$

is a vector subbundle of E which will be called the <u>image</u> of f.

Moreover, if E' is a vector subbundle of E, we can define a new vector bundle E'' over B by putting

$$E'' = \sum_{b \in B} E_b / E'_b \, .$$

In fact E'' is a vector bundle with fiber F/F' (F and F' being the fibers of E and E', respectively) and which will be called the <u>quotient</u> vector bundle of E by E'.

Let now (E,p,B), (E',p',B') and (E'',p'',B'') be vector bundles and $f\colon E \to E'$, $g\colon E' \to E''$ vector bundle homomorphisms over B. The sequence

$$E \xrightarrow{f} E' \xrightarrow{g} E''$$

is said to be <u>exact</u> if for each $b \in B$ the sequence of vector spaces

$$E_b \xrightarrow{f_b} E'_b \xrightarrow{g_b} E''_b$$

is exact. In such a case, we put

$$0 \to E \xrightarrow{f} E' \xrightarrow{g} E'' \to 0.$$

EXAMPLE. If E' is a vector subbundle of E, then the sequence

$$0 \to E \xrightarrow{f} E' \xrightarrow{g} E'' \to 0$$

is exact, where E'' is the quotient vector bundle of E' by E and g the canonical projection.

Next, we show that the algebraic operations on vector spaces can be extended to the vector bundles.

Indeed, let (E,p,B) be a vector bundle with fiber F and $\{U_i\}_{i\in I}$ an open covering of B with transition functions h_{ji}. Then we can define a vector bundle over B with fiber F^* (and of F) from the transition functions h^*_{ji} defined by

$$h^*_{ji}(b) = (h_{ji}(b)^*)^{-1}, \qquad b \in U_i \cap U_j ,$$

where $h_{ji}(b)^*$ is the adjoint of the linear mapping $h_{ji}(b)\colon F \to F$. This vector bundle E^* will be called the <u>dual</u> vector bundle of E.

REMARK. If $v \in E_b$ and $\alpha \in E^*_b$, E_b and E^*_b being the fibers of E and E' over $b \in B$, respectively, we can define a natural pairing $\langle v,\alpha\rangle$ as follows. Let h_i and h^*_i the local trivializations of E on E' over U_i, respectively, and suppose $b \in U_i$. Then we put $\langle v,\alpha\rangle = \langle v_i,\alpha_i\rangle$, where $h_i(v) = (b,v_i)$, $h^*_i(\alpha) = (b,\alpha_i)$, $v_i \in F$, $\alpha_i \in F^*$. Since $\langle v,\alpha\rangle$ does not depend on the choice of h_i and h^*_i, we deduce that E^*_b is precisely the dual vector space to E_b.

We can generalize the above construction as follows. Let $\Lambda^p F^*$ be the vector space of p-forms on F. Then we can define a vector bundle over B with fiber $\Lambda^p F^*$ and with transition functions given by

$$h^*_{ji}(b) = (h_{ji}(b)^*)^{-1}, \qquad b \in U_i \cap U_j ,$$

where $h_{ji}(b)^*: \Lambda^p F^* \to \Lambda^p F^*$ is now the automorphism of $\Lambda^p F^*$ induced by $h_{ji}(b)$. This vector bundle will be denoted by $\Lambda^p E^*$ and called the vector bundle of p-forms of E. Obviously, we have $\Lambda^1 E^* = E^*$. In particular, if E is the tangent bundle of a given manifold B, then $\Lambda^p T^* B$ is the vector bundle of p-forms on B and $\Lambda^p B$ is the set of sections of this vector bundle.

Now, let (E,p,B) and (E',p',B') be vector bundles over B with fibers F and F', respectively. We shall define a new vector bundle $E \oplus E'$ over B with fiber $F \oplus F'$ as follows. Let $\{U_i\}_{i\in I}$ and open covering of B and $h_{ji}: U_i \cap U_j \to GL(F)$, $h'_{ji}: U_i \cap U_j \to GL(F')$ the corresponding transition functions of E and E', respectively. Then the vector bundle $E \oplus E'$ has transitions functions

$$\begin{aligned} U_i \cap U_j &\to GL(F \oplus F') \\ b &\rightsquigarrow h_{ji}(b) \oplus h'_{ji}(b) \end{aligned}$$

and the fiber of $E \oplus E'$ over $b \in B$ is precisely $E_b \oplus E'_b$; the vector bundle $E \oplus E'$ is said to be the Whitney sum of E and E'.

The following proposition is immediate.

PROPOSITION (3). *Let (E,p,B) and (E',p',B') be vector bundles over B. Then E and E' are equivalent if and only if there exists a differentiable mapping $h: E \oplus E' \to R$ such that the restriction of h to any fiber of $E \oplus E'$ is a non degenerate bilinear form.*

APPENDIX B

CONSTRAINED SYSTEMS

B.1. - Introduction

It was Dirac (1950), (1964) who proposed a Hamiltonian theory for Lagrangians systems which cannot be directly put into this form, (see also Bergmann (1956)). This means that the condition on the non-singularity of the Hessian of the Lagrangian function with respect to the velocities breaks down. In such a case we say that the Lagrangian is degenerate, irregular or even, singular.

As we know, in order to obtain a Hamiltonian formalism in Analytical Mechanics we need all momentum variables to be independent of the velocity variables. If we drop the non-singularity condition then there are a certain number of momenta which are not entirely independent of the velocities - we may develop, of course, a Hamiltonian formalism including these momenta but this is against the usual procedure. Therefore we need to restrict our attention to a certain subset K of the phase space manifold for which the momenta are, as in the standard theory independent of the velocities. This subset is taken as a smooth manifold and is called constraint manifold (we recall the reader that constraints appear in Mechanics when the motion of the system is limited by initial relations: here the term constraint has a different meaning since it is related to the Lagrangian equations of motion).

Geometrically, Dirac constraint theory may be viewed as follows. Suppose that (W,ω) is an arbitrary symplectic manifold with a Hamiltonian $H: W \to R$. The canonical equations are given by the symplectic relation

$$i_X\omega = dH. \tag{1}$$

Considering the set of vector fields (resp. 1-forms) on W, $\chi(W)$ (resp. $\Lambda^1(W)$) we have that the linear mapping

$$s_\omega: \chi(W) \to \Lambda^1(W); \quad s_\omega(X) = i_X w,$$

is an isomorphism since ω is symplectic. Hence, for every 1-form v there is a unique vector field X_v solving equation (1), i.e., $i_{X_v}\omega = v$. We call X_v the <u>dual</u> of v with respect to the symplectic form ω. In particular, when $v = dH$ the dual vector field is called Hamiltonian and is represented simply by X_H.

Suppose that K is a manifold which can be regularly immersed - embedded - in W by a mapping $\varphi: K \to W$. Then putting

$$\varphi^*\omega = \bar{\omega} \tag{2}$$

we have that in general $(K,\bar{\omega})$ is not a symplectic manifold and $s_{\bar{\omega}}: \chi(K) \to \Lambda^1(K)$ is not an isomorphism. Our problem is:

$(i)_1$ Let h be a smooth function on K, embedded in (W,ω). What are the conditions for h to be extended into a smooth function H on W such that the corresponding Hamiltonian vector field X_H is tangent to K ? (This condition says that the motion of the system has to be constrained to lie entirely in K).

We can examine the following question, since we start with a symplectic manifold:

$(i)_2$ Given a Hamiltonian function H on a symplectic manifold (W,ω) and considering a manifold K embedded in W, under what conditions the equation

$$i_X\omega = dH \text{ restricted to } K \quad (\text{i.e., } i_X\bar{\omega} = d(\varphi^*H))$$

admits a solution X tangent to K at every point?

Recall that if H is an extension of h to all W then from the isomorphism s_ω one obtains a <u>unique</u> X_H solving the equation $i_X\omega = dH$, but such X_H may <u>not be tangent to</u> K (and so X_H is not an ordinary differential equation along K). The restriction of the dynamic equations to K does not assure us that there is a solution of the problem. We must have the <u>range condition</u> verified: $d(\varphi^*H) \in s_{\bar{\omega}}(\chi(K))$. In such a case, we cannot assure the unicity of X.

In general the solution of the problem leads to an algorithm since we need to find a more "restricted" constraint manifold along which the dual vector field is in fact a first order differential equation. For this reason we call the manifold K <u>primary constraint manifold</u> and the functions which locally define it <u>primary constraint functions</u>. The functions which define the "reduced" submanifolds of K are called <u>secondary</u>. Primary and secondary constraints are divised in two classes: the <u>first class</u> functions, which are in involution with respect to the Poisson brackets $\{\ ,\ \}$, i.e., $\{f,g\} = 0$ and the second class functions which are those which are not first class. We will see in the following a reason for such subdivision.

A geometrical approach for Constraint theory was proposed by Gotay and co-workers (Gotay, Nester and Hinds (1978)) generalizing in an intrinsical way the Dirac theory for infinite dimensional manifolds. The theory was taken in a more general context since we may replace the symplectic form ω by a presymplectic form Ω. Then the induced linear mapping $s_\Omega: \chi(W) \to \Lambda^1(W)$ is not surjective (see Chapter II, §2.2).

We may adopt a point of view inspired in Lichnerowicz (1975): in place of searching a final constraint manifold K (if it exists) solving the problem, we study the class of functions for which the vector fields are 1st order differential equations along K fixed. From such point of view it is possible to develop a formalism inspired in Thom's Catastrophe theory for a class of constraints satisfying some especific properties. This was shown by Pnevmatikos ((1979)(a), (1979)(b),(1983)).

Constraint theory may also be formulated directly for Lagrangians (Shanmugadhasan (1963),(1973)). If we develop the Euler-Lagrange equations for a Lagrangian $L: TN \to \mathbb{R}$, N a given manifold of dimension n, then one obtains an expression of type

$$V_{ij}\,\ddot{q}^j = f_i\,, \tag{3}$$

where $V_{ij} = \dfrac{\partial^2 L}{\partial \dot{q}^i \partial \dot{q}^j}$ and f_i stands for the remaining terms.

The degeneracy of the matrix (V_{ij}) says that these equations cannot be solved uniquely for $\ddot{q}^j$. If the rank of (V_{ij}) is r, $r < n$, then $n-r$ Lagrange equations are of order less than two. Therefore we have a set of constraints to be included in the study and similarly to the Hamiltonian situation other constraints may appear in the theory. A discussion about the nature of such constraints (with respect to the Hamiltonian ones) is pertinent (see Gotay & Nester (1979) and Schafir (1982)). Of course a geometrical approach may also be developed and we will see this very briefly in the present Appendix. We will summarize some results of Gotay et al and Pneumatikos and indicate their articles for more details.

B.2. - Presymplectic systems

2.1. Singular sets.

The symplectic description of both (Lagrangian-Hamiltonian) problems may be summed up as follows: W is a $2n$-dimensional manifold which may be considered the tangent (for Lagrangians) or cotangent (for Hamiltonians) bundle of a given n-dimensional manifold N. Let w be a symplectic form on W. If the formalism is Lagrangian then w is supposed to be $-dd_J L$, where J is the almost tangent structure on TN and L is supposed regular. If the formalism is Hamiltonian then w is the symplectic form canonically obtained from the Liouville form on T^*N. As we have reminded above, a more general context is given when we consider (W,w) presymplectic. In such a case, for the Lagrangian theory we do not impose the regularity condition on L and in the Hamiltonian theory w is a closed two form not necessarily of maximal rank. In the present work in general we will suppose that w is symplectic.

The Lagrangian and Hamiltonian equations of motion can be symbolically written as

$$i_X w = dE, \tag{4}$$

where E is a smooth function on W.

Let K be a manifold embedded into W by a mapping φ and let $\bar{w}$ be as in (2). Since $(K,\bar{w})$ will not be in general a symplectic manifold we may consider the sets

$$S_c\bar{w} = \{z \in K;\ \operatorname{corank}_z \bar{w} = c\}, \qquad 0 \le c \le \dim K.$$

Recall that the rank (resp. the corank) of a 2-form v on a manifold A in a point $z \in A$ is the rank (resp. the corank) of the linear mapping $s_v\colon T_zA \to T_z^*A$, $s_v(X(z)) = i_{X(z)}\, v(z)$.

The <u>singular set</u> $S_{\bar{w}} = \bigcup_c S_c\bar{w}$ is an obstruction for K to be a symplectic manifold (the parity of the dimension of K will also be a problem). If the dimension of K is even and if for every z, $\operatorname{corank}_z \bar{w} = 0$, then the embedding φ is called symplectic: otherwise it is called presymplectic and $(K,\bar{w})$ is said to be a <u>presymplectic constraint manifold</u>. In the following we will identify K with $\varphi(K)$, TK with $T\varphi(K)$, etc., for simplicity.

REMARK (1). We have mentioned before that we look for an X satisfying (4) such that X is tangent to K everywhere, i.e., X is a first order differential equation on K. An important difference between Lagrangian and Hamiltonian formalisms with constraints is that for the first one we need that X also be a second order differential equation on K. This is due to the fact that the integrals of X need to be solutions of the second order Euler-Lagrange differential equations of motion.

The geometrical formulation of Gotay et al is developed by supposing that for every $z \in K$, $\operatorname{corank}_z \bar{w} = \text{constant} \neq 0$, that is, $S\bar{w} = K$. The general situation will be seen in Pnevmatikos approach.

Suppose that (P,v) is a presymplectic manifold (in the sense that s_v is not an isomorphism). Consider the equation

$$u_X v = u, \tag{5}$$

where u is a 1-form on P. If such form is in the range of s_v then the equation can be solved and $X(z) \in T_zP$ for every $z \in P$ since we are not taking P as an embedded manifold. But in general there are only some points z in P for which $u(z)$ is in the range of s_v. Assuming that the set of such points form a manifold P_1 we embed P_1 into P by a regular immersion (if this is possible) and we consider now the equation on P_1 given by

$$\varphi_1^*(i_X v - u) = 0,$$

with u in $s_v(\varphi^* TP_1)$. The solution X may be or not tangent to P_1. If it is, the problem is solved. If it is not then we restrict our attention to a second manifold P_2; we examine again the range and the tangency conditions and so on.

It is clear that the repetition of such arguments generates a sequence of manifolds

$$\cdots \to P_{i+1} \to P_i \to \cdots P_1 \to P$$

where

$$P_{i+1} = \{z \in P_i \, ; \; u(z) \text{ is in the range of } s_v(TP_i)\}$$

(where for simplicity we have put s_v for $s_{\varphi_i^* \circ \cdots \circ \varphi_1^*(v)}$).

This process gives the following possibilities.

(i) The algorithm preduces an integer r such that $P_r = \emptyset$;

(ii) The algorithm produces an integer r such that $P_r \neq \emptyset$ but $\dim P_r = 0$;

(iii) There is an integer r such that $P_r = P_t$, for all $t \geq r$ and $\dim P_r \neq 0$;

(iv) The process is not finite.

The first case means that dynamical equations have no solutions. Case (ii) says that the only solution is $X \equiv 0$ and so there is no dynamics. The third possibility gives the <u>final</u> constraint manifold in the sense that if N is any other submanifold along which the dynamical equations are verified then $N \subset P_r$. Case (iv) says that the system has an infinite degree of freedom. In such a case we may stipulate that if P_∞ is the intersection of all P_r then we may reproduce one of the other possibilities.

Clearly, our interest here is the situation described by (iii).

2.2. Conditions to solve the algorithm.

The above description is unsatisfactory since it only shows how the problem is pictured without saying how is it directly solved. We are interested here only in the first order problem and so we make no distinction between the Hamiltonian and Lagragian aspects. Since, the problem starts when the range conditions is verified we suppose also that such condition is initially verified.

Suppose that (W,w) is a symplectic manifold and $(K,\bar{w})$ is a constraint manifold veryfing the range condition, embedded in W by a mapping $\varphi: K \to W$. We may suppose that an initial function h is given on K in the sense that h corresponds to the regular part of the theory as we have remarked in the introduction. In other words, we initiate our study with some 1-form u and we suppose that $u = dh$ along the constraint K, where h is globally defined on K for the regular part of the Lagrangian. We wish to know how we can solve the first order problem.

Let (x,y) be a coordinate system defined on a neighborhood $V \subset W$ of some point z in K such that $y = y^B = (y^{r+1},\ldots,y^{2n})/U = K \cap W \equiv 0$, with $\dim K = r$ and $\dim W = 2n$. We extend h to V by

$$H = h + u_B y^B \tag{6}$$

with u_B being arbitrary multipliers to be determined. Expression (6) gives the local conditions to extend symplectically the Hamiltonian h and the restriction of the equation

$$i_X w = dH \tag{7$_1$}$$

to K gives

$$i_X \bar{w} = dh\,. \tag{7$_2$}$$

Equation $(7)_1$ has of course a solution X_H defined for all points in V and in particular for those in U. X_H is tangent to V but not necessarily to U. As we have supposed that $dh \in s_{\bar{w}}(\chi(U))$ our main problem is the tangency in the first order problem. Of course we suppose also that the extension H is a smooth function on W.

Let z be a point in K and consider the symplectic complement $T_zK^{\perp}$ of T_zK in (T_zW, w_z):

$$T_zK^{\perp} = \{X \in T_xW;\quad \text{for all}\quad Y \in T_zK;\quad i_Xw(z)Y = 0\}.$$

Locally the symplectic complement $T_zV^{\perp}$ is generated by the family (X_{y^B}) of dual vector fields of dy^B. In the following we represent X_{y^B} by X_B, $r+1 \le B \le 2n$. If we denote by $C^{\infty}(W)$ the set of differentiable functions on W, we can consider the following set

$$K(W,w) = \{F \in C^{\infty}(W);\quad \text{for all}\quad z \in K,\quad X_F(z) \in T_zK\},$$

(recall that X_F is such that $i_{X_F} w = dF$).

PROPOSITION (1). $K(W,w) = \{F \in C^{\infty}(W);$ for all $z \in K$ and $X \in T_zK^{\perp}$, $dF(X) = X(F) = 0\}$.

Proof: It is sufficient to show the proposition for the family X_B. If F is a smooth function on W then X_F is tangent to K in a neighborhood of a point z if and only if $X_F(y^B) = 0$ since for an integral curve $f(t)$ of X_F

passing by z we have $X_F(f(t)) = (d/dt)(f(t))|_{t=0}$. The local characterization of K by the functions y^B tells us that all y's are constant along the trajectories of X_F and so $X_F(y^B(f(t))) = (d/dt)(y^B(f(t))) = 0$. But

$$X_F(y^B) = -X_B(F) = -dF(X_B)$$

which gives the above characterization of $K(W,w)$. □

DEFINITION (1). The functions $F \in K(W,w)$ are called <u>first class</u>. Functions which are not first class are called <u>second class</u>.

First class functions may be defined with respect to the Poisson brackets. If (x,y) is a local coordinate system in an open neighborhood V of W such that $y = (y^A) = 0$ along $U = V \cap K$, then F is a first class function if for all A one has $\{F,y^A\} = 0$. In fact

$$\{F,y^A\} = w(X_F,X_A) = X_A(F) = dF(X_A).$$

We will say that (y^A) are <u>first class constraint functions</u> if $y^A \in K(W,w)$ for all A. <u>A first class constraint manifold</u> of (W,w) is a submanifold K of W characterized locally by the vanishing of a set of functions which are in $K(W,w)$.

Consider again a local coordinate system $(x,y) = (x_i,y^A) = (x_1,\ldots,x_r,y^{r+1},\ldots,y^{2n})$ on V for which $U = V \cap K$ is defined by $y = 0$. If $\{(\partial/\partial x_i),(\partial/\partial y^A)\}$ is the basis induced by such coordinates then the restriction to U of $(\partial/\partial y^A)$ generates a subbundle S of $TW|U$ supplementary of TU in $TW|U$. The dual vector fields X_A of dy^A with respect to the symplectic form w generates a subbundle $TU^{\perp}$ of $TW|U$ and therefore we may put

$$X_A = T_A + S_A$$

for the direct decomposition $TW|U = TU \oplus S$, where T_A and S_A are sections of TU and S, respectively. A direct calculation shows that

$$X_A(H) = T_A(h) + S_A(H) \tag{8}$$

and $X_A(H) = 0$ if and only if $T_A(h) = -S_A(H)$. Taking the differential of (6) we see directly that

$$dH(X_A) = X_A(H) = dh(X_A) + u_B dy^B(X_A)/U \Rightarrow \begin{cases} X_A(H) = T_A(h) + u_B S_A(y^B) & (9)_1 \\ X_A(H) = T_A(h) + u_B\{y^A,y^B\} & (9)_2 \end{cases}$$

and

$$S_A(H) = u_B S_A(y^B) = u_B\{y^A,y^B\}. \tag{10}$$

PROPOSITION (2). <u>Let</u> (x,y) <u>be a coordinate system as above. A necessary and sufficient condition for a smooth function</u> h <u>on</u> $(K,\bar{w})$ <u>to be extended along a neighborhood</u> V,

$$H = h + u_B y^B, \qquad r+1 \le B \le 2n$$

<u>with the property that the dual vector field</u> X_H <u>is tangent to</u> K <u>at every point in</u> U <u>is that the system</u>

$$u_B\{y^A,y^B\} = -T_A(h) \tag{11}$$

<u>admits a solution</u> $u = (u_B)$ <u>for such coordinates</u>.

Proof: If (11) is verified then from $(9)_2$ we have that for every vector field $Z \in TU^{\perp}$, $Z(H) = 0$ and from Proposition (1), X_H is tangent to U.

If H is in $K(W,w)$ then $X_A(H) = 0$ and so $S_A(H) = -T_A(h)$. Taking into account the coordinates (x,y), we develop S_A along the basis $(\partial/\partial y^A)/U$ and we obtain

$$-T_A(h) = \{y^A, y^B\}(\partial H/\partial y^B)/U.$$

Therefore the functions $(\partial H/\partial y^B)/U$ are the desired solutions. □

When are there such coordinate systems? If for a given system the matrix $\pi = (\{y^A, y^B\})$ is of maximal rank then of course our problem is solvable. But in general this will not be so. We must discard the case where (11) is insoluble for every coordinate system. In such a case Dirac says: "it would mean that the Lagrangian equations of motion are inconsistent and we are excluding such a case." (Dirac (1964)).

Suppose that for a given coordinate system the rank of $\pi = (\{y^A, y^B\})$ is not maximal. Since $T_A(h)$ in (11) is arbitrarily chosen it may occur that all components of such a vector field are not in the image of the regular part of π. Ours system is independent of the u's and $X(H) = 0$ if $T_A(h) = 0$. This equation will not be satisfied except in an open subset U_1 of U. This subset is characterized by a coordinate system $(x_a, \bar{y}^\alpha)$, where now $1 \le a \le k$, $k+1 \le \alpha \le r$.

Starting again with $X_\alpha(H) = T_\alpha(h) + S_\alpha(H)$ for such situation, we analyse the system

$$- T_\alpha(h) = u_B\{\bar{y}^\alpha, y^B\}$$

where $S_\alpha(H) = u_B\{\bar{y}^\alpha, y^B\}$ is obtained from $dH(X_\alpha) = dh(X_\alpha) + u_B y^B(X_\alpha)$. If the above system is independent of the u's then we continue the process into another subset U_2 of U_1 and so on. The iteration procedure originates a final local constraint submanifold U_k for which at least some of the components of the corresponding $T(h)$ are in the image of the regular part of the respective matrix constructed for such domain.

We suppose, for simplicity, that for our original system (x_i, y^A), s-elements of $T_A(h)$ are in the image of the regular part of π, where s is the rank of π. In such a case we know from Linear Algebra that the system (11) has a solution of type

$$u_B = \bar{u}_B + v_\alpha \bar{u}_{\alpha B}$$

where $\bar{u}_B$ is fixed and $v_\alpha \bar{u}_{\alpha B}$ is a linear combination of all solutions of the homogeneous equations associated to (11). If we put

$$f^\alpha = \bar{u}_{\alpha B}\, y^B$$

then

$$H = h + \bar{u}_B y^B + v_\alpha f^\alpha$$

is the final expression of the extended Hamiltonian. We can easily verify that H is of first class as well as $h + \bar{u}_B y^B$ and f^α. Finally, it is clear that, in the general situation, at the end of the process our Hamiltonian will be of type

H = h + first class (primary/secondary) constraints +
+ second class (primary/secondary) constraints.

Constraint theory may be also formulated for Lagrangians starting with a regular function $L: W \to \mathbb{R}$, where now $W = TN$ is the tangent bundle of some configuration manifold N of dimension n and w is the symplectic form $-dd_J L$.

The symplectic version of the Euler-Lagrange equations in Hamiltonian form is

$$i_X w = dE \qquad (*)$$

where E is the energy defined by $C(L)-L$ and C is the Liouville vector field.

We examine (*) in an analogous way as in the Hamiltonian procedure. However, as we have remarked before, the constraint algorithm is insufficient to entirely solve the problem since it does not assure us that the solution X is naturally a second order differential equation along the final constraint manifold. Therefore we incorporate to the problem the equation

$$JX = C \qquad (**)$$

restricted to the final constraint manifold, (we recall the reader that (**) gives the 2nd order condition for vector fields - see Chapter II).

Gotay and Nester ((1979)(b)) studied such a problem. Their main result is given for a special class of Lagrangian functions. Let $W = TN$, $w = -dd_J L$ and $V(TN)$ the vertical vector fields with respect to the projection $P_{T_N}: T(TN) \to TN$.

DEFINITION (2). Let $L: W \to \mathbb{R}$ be a smooth function. We say that (W,L,w) is an <u>admissible system</u> if the involutive distribution $D = \operatorname{Ker} w \cap V(TN)$ generates a foliation $\mathfrak{F}$ on W

such that the leaf space $\hat{W} = W/\mathfrak{F}$ admits a manifold structure of class C^∞ such that the canonical projection $\hat{p}: W \to \hat{W}$ is a submanifold. In such a case we say that L is an admissible Lagrangian.

Roughly speaking, a foliation of dimension r on a manifold is a partition of M into disjoint connected subsets $\{F_a\}_a$ characterized locally by coordinate systems $(x,y): U \to \mathbb{R}^r \times \mathbb{R}^s$ such that for each F_a, $U \cap F_a$ has local equations $y_1 = \text{constant}, \ldots, y_s = \text{constant}$. The subsets F_a are called the leaves of the foliation which inherits the topology of M only if they are embedded in M. The foliation $\{F_a\}_a$ is represented by $\mathfrak{F}$ and introducing the equivalent relation R which states that xRy if they are in the same leaf, we define the leaf space $\hat{M} = M/\mathfrak{F}$ by the quotient of M by $\mathfrak{F}$. For more details on the theory of foliations see Blaine Lawson (1977), for example.

THEOREM (1) (The second order equation theorem). Let (W,L,w) be an admissible Lagrangian system with final constraint manifold K embedded in W. Then there exists at least one submanifold P of K and a unique smooth vector field X on P (for fixed P) satisfying simultaneously the equations

$$i_X w - dE/P = 0 \quad \text{and} \quad JX - V/P = 0.$$

Every such manifold P is diffeomorphic to the leaf space $\hat{K} = K/\mathfrak{F}_K$, where $\mathfrak{F}_K = \mathfrak{F}/K$ is the induced foliation on K obtained from $\mathfrak{F}$.

(For a proof see Gotay & Nester (1979)(b)).

B.3. - Generic constraints

3.1. Preliminaries.

The singular set $S\bar{w}$ was so forth supposed to be K. We consider now K decomposed by the sets $S_c\bar{w}$ defined before, where c is an integer varying continuously. A study of this situation was proposed by S. Pnevmatikos using some ideas and techniques of Singular theory of differential forms. We summarize now the principal results of such point of view and we indicate the article of Martinet (1970) as well as the book of Golubtsky and Guillemin (1973) for the main properties and definitions on the subject.

In the following the space of all differential forms of some finite degree with smooth coefficients defined on a given manifold is endowed with the Whitney C^∞-topology (see Golubtsky & Guillemin (1973), p.42). They form a Baire space under such topology.

DEFINITION (3). A stratification on a smooth manifold is a partition $S = (S_i)_i$ of M by a family of subsets S_i of M such that each S_i (called stratum) is a smooth submanifold of M and for each integer j, the union of the strata with dimension $\leq j$ is a closed subset in the topology of M. The stratification is said to be coherent if for every z in M and every vector v in T_zM there is an open set V, $z \in V$ and a smooth vector field X on V such that $X(z) = v$ and $X(x) \in T_xS_i \subset T_xM$, where S_i is the stratum containing z.

Suppose that E is an n-dimensional vector space and $v \in \Lambda^2 E^*$ is a two form. Let $r = \text{rank } v < n$ and put $c = n-r$. Then the sets

$$\Sigma_c = \{v \in \Lambda^2 E^*;\ \text{corank } v = c\}$$

are submanifolds of $\Lambda^2 E^*$ with codimension $c(c-1)/2$ (cf. Martinet (1970)). Further, these manifolds form a coherent stratification of $\Lambda^2 E^*$ (this result, of course, may be extended to forms with degree higher than 2).

On the other hand, as the rank of every 2-form is even we have: $\Sigma_c \neq \emptyset$ if and only if $n-c$ is even. Hence, if

n is even and $c = 0,2,4,6,\ldots$, then $\text{cod } \Sigma_c = 0,1,6,15,\ldots$

n is odd and $c = 1,3,5,7,\ldots$, then $\text{cod } \Sigma_c = 0,3,10,21,\ldots$

Let K be a smooth manifold of dimension k. The bundle $\Lambda^2(T^*K)$ has the space $\Lambda^2(R^k)^*$ as fibre. It can be shown that the stratification of $\Lambda^2(R^k)^*$ by manifolds Σ_c induces a stratification on $\Lambda^2(T^*K)$ by smooth manifolds $S_c(K)$ which are fibred over K with fibre type Σ_c. We represent by $S(K)$ such stratification of $\Lambda^2(T^*K)$ and it is clear that

$$S_c(K) = \{v \in \Lambda^2(T^*K);\ \text{corank } v = c\}.$$

DEFINITION (4). Consider two smooth manifolds A and B. A smooth mapping $f: A \to B$ is said to be <u>transversal</u> (or <u>intersects transvesally</u>) a subset C of B in a point $a \in A$ such that $f(a) \in C$ if the following decomposition is verified

$$Tf(a)(T_a A) + T_{f(a)}C = T_{f(a)}B\ .$$

We say that f is transversal to C if it is transversal to C at every point of $a \in A$ with $f(a) \in C$. If S is a stratification of B then f is transversal to S if it is transversal to each stratum S_c of S.

From Thom Transversality Theorem (Golubitsky and Guillemin (1973), p.54) it can be shown that the set $\Lambda^2_{\pitchfork} K$ of all 2-forms on K which intersects $S(\kappa)$ transversally is residual, that is, this set is the countable intersection of open dense subsets of $\Lambda^2 K$. Since $\Lambda^2 K$ is a Baire space in the Whitney C^∞-topology, $\Lambda^2_{\psi} K$ is an open dense set of $\Lambda^2 K$ and every open subset of $\Lambda^2 K$ has non-emtpy intersection with $\Lambda^2_{\pitchfork} K$.

3.2. Generic embeddings and consistent Hamiltonians.

The above considerations suggest us the following

DEFINITION (5). A two form $\omega \in \Lambda^2 K$ is said to be generic over $\Lambda^2 K$ if it is transversal to the stratification $S(k)$ of $\Lambda^2(T^*K)$.

Let (W,w) be a symplectic manifold, K a smooth manifold of dimension r, $r \leq 2n = \dim W$ and $\mathcal{E}(K,W)$ the set of all smooth embeddings from K into W. A property of embeddings of K into W is said generic on $\mathcal{E}(K,W)$ if this property is verified on an enumerable intersection of open and dense subsets of $\mathcal{E}(K,W)$. Prnevmatikos showed that

> "the restriction $\bar{w} = i^*w$ of the symplectic form w to K by the embedding $i: K \to M$ is transverse to the stratification by the rank $S(K)$ of $\Lambda^2(T^*K)$,"

is a generic property of embeddings.

Roughly speaking we may say that an embedding i is generic if the form $\bar{w}$ is generic over $\Lambda^2 K$, but we need to give an explicit definition of genericity for embeddings since this notion is related to transversality. The definition of transversality is strictly dependent of the 1-jet of the considered mapping and therefore we say:

$i \in \mathcal{E}(K,W)$ is generic if the one-jet $\tilde{i}^1$ is transversal to the stratification of the one-jets of embeddings $J^1(\mathcal{E}(K,W))$ by the submanifolds

$$K_c = \varphi^{-1}(S_c K) \cap J^1(\mathcal{E}(K,W)),$$

where φ is the submersion $\tilde{i}^1(x) \mapsto i^* w(x)$ from $J^1(K,W)$ to $\Lambda^2(T^* K)$. It can be shown that this definition is equivalent to Definition (5) (see Pnevmatikos (1983)).

Let v be a 2-form on K. A point $z \in K$ is <u>singular</u> (resp. <u>regular</u>) point of v if $\operatorname{corank}_z v \neq 0$ (resp. $\operatorname{corank}_z v = 0$). The singular points of v generate a singular set $S_v = \{z \in K;\ \operatorname{corank}_z v \neq 0\}$ and in particular the sets

$$S_c v = \{z \in K;\ \operatorname{corank}_z v = c\} \qquad 0 \leq c \leq \dim K.$$

It can be shown that if v is generic then the family $S_v = (S_c v)_c$ is a coherent stratification of K where the codimension of each $S_c v$ is $c(c-1)/2$. Therefore we can see that along the manifold K is defined a coherent stratification for the form $\bar{w} = i^* w$, where i is a generic embedding of K into W. It can be shown that $S_{\bar{w}}$ is finite for different generic embeddings.

DEFINITION (6). Suppose $i \in \mathcal{E}(K,W)$ be generic. Then we say that $(K,\bar{w})$ is a generic (presymplectic) constraint manifold.

DEFINITION (7). Let $(K,\bar{w})$ be a presymplectic manifold. A smooth function h on K is said to be a consistent Hamiltonian for $(K,\bar{w})$ if there is a unique vector field X_h such that $i_{X_h}\bar{w} = dh$. We say that h satisfies the kernel condition with respect to $\bar{w}$ if $\mathrm{Ker}\, dh(z) \supset \mathrm{Ker}\, \bar{w}(z)$ for all $z \in K$. If $(K,\bar{w})$ is a presymplectic constraint manifold of a symplectic manifold (W,w), $\bar{w} = i^*w$, then we say that h admits a symplectic extension on (W,w) if there is a smooth extension $H: W \to \mathbb{R}$ of h such that X_H is tangent to K.

The following theorem is due to Pnevmatikos:

THEOREM (2). If $(K,\bar{w})$ is a generic constraint manifold of the symplectic manifold (W,w) such that the dimension of K is even then the following statements are equivalent:

(i) h is a consistent Hamiltonian for $(K,\bar{w})$

(ii) h admits a symplectic extension on (W,w)

(iii) h satisfies the kernel condition at almost any point of $S\bar{w}$.

The proof is, of course, not simple. It presupposes for example a good knowledge of the Singular theory of functions and forms. An essential result, obtained from this study, consists in the demonstration that for generic constraints equation (11) in Proposition (2) is always solvable when the kernel condition is verified. As such a study is out of the purposes of the present book we suggest to the reader the article of Pnevmatikos (1983) (as well as the references therein) for a detailled exposition on the subject.

The above theorem gives a characterization on consistent Hamiltonians in the following way:

COROLLARY. *Let* $(K,\bar{w})$ *be a generic constraint manifold of* (W,w) *of even dimension. Then the set of regular consistent Hamiltonians on* K *are precisely those functions which satisfy the kernel condition with respect to* $\bar{w}$.

REMARK (2). If $\dim K$ is odd then it is possible to establish similar results (cf. Pnevmatikos (1983)). In fact this situation corresponds to non-autonomous Mechanics and we work with contact forms in the place of symplectic forms. If K is a manifold of dimension $2k-1$ and $\Omega = u \wedge dt + v \in \Lambda^2(K \times R)$, where $(u,v) \in \Lambda^1(K) \times \Lambda^2(K)$ then we want to know what are the necessary and sufficient conditions for the equation

$$i_{X+f\frac{\partial}{\partial t}} \Omega = \alpha + g dt$$

to be solvable with an (eventually) unique solution $X \in \chi(K)$ and $f \in C^\infty(K)$ and $\alpha \in \Lambda^1(K)$. In particular one has $\Omega = -dH \wedge dt + w$, where w is a symplectic form on W.

REMARK (3). The unicity of the solution is obtained from the regular part of K. When we work with a presymplectic manifold without embedding it in a symplectic manifold, the vector field (if exists) which solves the constraint dynamic equations is not unique since we may add to it any vector field which is in the kernel of the presymplectic form restricted to the final constraint manifold. If the presymplectic manifold is embedded generically into a symplectic manifold then the original Hamiltonian h, satisfying the kernel condition, admits a symplectic extension H (not necessarily unique) and putting $X_h = X_H/K$ we see that the unicity of X_h along the regular set $S_o\bar{w}$ gives the desired result along K. Let us remark that X_h is tangent to K at every point but it is not necessarily tangent to each stratum of K.

Finally, we may try to develop a Lagrangian formalism for generic constraints. An attempt in this direction may be found in Rodrigues (1984).

REFERENCES

ABRAHAM, R. & MARSDEN, J. - (1978), Foundations of Mechanics, 2nd ed., Benjamin, N.Y.

AKHIEZER, A. & BERETETSKY, V. - (1965), Quantum Electrodynamics, Interscience Publ., N.Y.

ALDAYA, V. & de AZCÁRRAGA, J. -

(1978)(a), Vector bundles, r^{th} order Noether invariants and canonical symmetries in Lagrangian field theory, J. Math. Phys., 19, (7), 1876-1880.

(1978)(b), Variational principles on r^{th} order jets of fibre bundles in field theory, J. Math. Phys., 19, (9), 1869-1875.

(1980), Geometric formulation of Classical Mechanics and field theory, Rev. Nuovo Cimento, 3, (3), (10), 1-66.

(1982), Higher order Hamiltonian formalism in field theory, J. Phys. A Math., 13, 2545-2551.

AMBROSE, W., PALAIS, R. & SINGER, W. - (1960), Sprays, Anais Acad. Bras. Cienc., 32, 163-178.

ANDERSON, D.

(1973)(a), Noether's theorem in Generalized Mechanics, J. Phys. A, 6, 299-305.

(1973)(b), Averaged Variational principles containing higher derivatives, Lett. Nuovo Cimento, 7, (9), 317-320.

(1973)(c), Noether's theorem and Variational integrals containing linear operators, Lett. Nuovo Cimento, 6, (9), 303-313.

(1973)(d), Equivalent Lagrangians in Generalized Mechanics, J. Math. Phys., 14, (7), 934-936.

(1983)(e), Averaged Lagrangians containing higher derivatives, J. Phys. A, (6), 1129-1139.

ANDERSON, D. & ASKNE, J. - (1973), An extension of Generalized Mechanics, Lett. Nuovo Cimento, 7, (13), 550-552.

ARENS, R.

(1981)(a), Reducing the order of a Lagrangian, Pacific J. Math., 6, 1-11.

(1981)(b), Manifestly Dynamic forms in the Cartan-Hamilton treatment of Classical fields, Pacif. J. Math., 6, 13-30.

ARNOLD, V. - (1974), Méthodes Mathématiques de la Mécanique Classique, Ed. MIR, Moscou.

BARTH, N. & CHRISTENSEN, S. - (1983), Quantizing four order gravity theories: the functional integral, Phys. Rev. D, 28, (8), 1876-1893.

BARUT, A. - (1980), Electrodynamics & Classical theory of fields and particles, Dover, N.Y.

BARUT, A. & MULLEN, G.

(1962)(a), Quantization of two-component higher order spinor equations, Annals of Phys., 20, 181-202.

(1962)(b), Action principle for higher order Lagrangians with an indefinite metric, Annals of Phys., 20, 203-218.

BERGER, M. - (1965), Lectures on geodesic in Riemannian geometry, Tata Institute, Bombay.

BOGOLIUBOV, N. & SHIRKOV, D. - (1959), Introduction to the theory of Quantized fields, Wiley & Sons, N.Y.

BOPP, F. - (1940), Eine lineare theorie des Elektrons, Ann. Physik, 38, 345-384.

BORNEAS, M.

(1959), On a generalization of the Lagrange function, Amer. J. Phys.

(1960), The Lagrange function in a general problem, Nuovo Cimento, (16), (5), 806-810.

(1963), On Lagrangians with high derivatives, Acta Phys. Polonia, 4, (10), 471-475.

(1969), Principle of Action with higher derivatives, Phys. Review, 186, (5), 1299-1303.

(1972)(a), A quantum equation of motion with higher derivatives, Amer. J. Phys., 40, 248-251.

(1972)(b), Some solutions of the Quantum Wave Equation with higher derivatives, Bul. St. Th. Trist. Pol., Timisoara, 17, (31), (2), 11-16.

BRICKELL, F. & CLARK, R.S. - (1970), Differentiable Manifolds, Van Nostrand, London.

CANDOTTI, E., PALMIERI, C. & VITALE, B.

(1970), On the inversion of Noether's theorem in the Lagrangian formalism, Nuovo Cimento, 70A, (2), 233-246.

(1972)(a), On the inversion of Noether's theorem in Classical Dynamical Systems, Amer. J. Phys., 40, 424-429.

(1972)(b), Universal Noether's Nature of infinitesimal transformations in Lorentz-covariant field theories, Nuovo Cimento, 7A, (1), 271-279.

CARTAN, E. - (1921), Leçons sur les invariants intégraux, Hermann, Paris.

CATZ, G. -

(1974)(a), Sur le fibré tangent d'ordre 2, C.R. Acad. Sc. Paris, 278 A, 277-278.

(1974)(b), Gerbes et connexions sur le fibré tangent d'ordre 2, C.R. Acad. Sc. Paris, 278 A, 347-349.

CHANG, T. -

(1946), A note on the Hamiltonian equations of motion, Proc. Camb. Phil. Soc. 42, 132-138.

(1948), Field theories with high derivatives, Proc. Camb. Phil. Soc. 44, 76-86.

CLARK, R. & BRUCKHEIMER, M. - (1960), Sur les structures presque tangentes, C.R. Acad. Sc. Paris, 251, 627-629.

CLARK, R. & GOEL, D. -

(1972)(a), On the gometry of an almost tangent manifold, Tensor, N.S. 24, 243-252.

(1972)(b), Almost tangent manifolds of 2nd order, Tohoku Math. J., 24, 79-92.

COELHO DE SOUZA, L. & RODRIGUES, P.R. - (1969), Field theory with higher derivatives - Hamiltonian structure, J. Phys. A 2, (2), 304-310.

DAZORD, P. - (1966), Sur une généralisation de la notion de spray, C.R. Acad. Sc. Paris, 263, 543-546.

de BARROS, C. -

(1964), Variétés hor-sympléctiques, C.R. Acad. Sc. Paris, 259, 1291-1294.

(1965), Variétés presque hor-complexes, C.R. Acad. Sc. Paris, 260, 1543-1546.

(1967), Sur la géométrie différentielle des formes différentielles extérieures quadratiques, Atti Congr. Int. Geometria Differenziale, Bologna, 1-26.

(1975), Systèmes Mécaniques sur une variété Banachique, C.R. Acad. Sc. Paris, 280, 1017-1020.

DEBEVER, R. - (1951), Les espaces de l'électromagnetisme, Colloque de Géométrie Différentielle, Louvain, Masson, Paris, 217-233.

DEDECKER, P. -

(1950), Sur les intégrales multiples du calcul des variations, C.R. IIIe Congrès National des Sciences, Bruxelles.

(1953), Calcul des Variations, formes différentielles et champs géodesiques, Colloque Int. de Géométrie Différentielle, Strasbourg.

(1977)(a), On the generalization of symplectic geometry to multiple integrals in the Calculus of Variations, Lect. N. Math. 570, 395-456.

(1977)(b), Généralisation d'une formule de H.A. Schwarz relative aux surfaces minima, C.R. Acad. Sc. Paris, 285 A, 23-26.

(1977)(c), Généralisation d'une formule de H.A.Schwarz aux intégrales multiple du Calcul des Variations, C.R. Acad. Sc. Paris, 285 A, 59-61.

(1977)(d), Intégrales complètes de l'équation aux dérivées partielles de Hamilton-Jacobi d'une intégral multiple, C.R. Acad. Sc. Paris, 285 A, 123-126.

(1978), Problèmes variationnels dégénérés, C.R. Acad. Sc. Paris, 286 A, 547-550.

(1979), Le Théorème de Helmhotz-Cartan pour une intégrale simple d'ordre supérieur, C.R. Acad. Sc. Paris, 288 A, 827-829.

(1984)(a), Existe-t-il, en Calcul des Variations, un formalisme de Hamilton-Jacobi-E. Cartan pour les intégrales multiples d'ordre supérieur?, C.R. Acad. Sc. Paris, 298 I, (16), 397-400.

(1984)(b), Sur le formalisme de Hamilton-Jacobi-E.Cartan pour une intégrale multiple d'ordre supérieur, Preprint.

de DONDER, Th. - (1935), Théorie invariantive du Calcul des Variations, Gauthier-Villars, Paris, 95-108.

de LEÓN, M. -

(1978), Conexiones y estruturas polinómicas en el fibrado tangente de orden 2, (Doctoral thesis), Publ. Dpto. Geom. y Topo., 48, Santiago de Compostela, 1-264.

(1981), Connections and f-structures on T^2M, Kodai Math. J., 4, 189-216.

(1982)(a), Systèmes Lagrangiens réguliers d'ordre supérieur, C.R. Acad. Sc. Paris, 294, 451-453.

(1982)(b), Transformation de Legendre pour les systèmes lagrangiens d'ordre supérieur, C.R. Acad. Sc. Paris, 295, 123-125.

de LEÓN, M. & VILLAVERDE, C. -

(1981)(a), Calcul différentiel sur les fibrés tangents d'ordre supérieur, C.R. Acad. Sc. Paris, 292 I, 881-884.

(1981)(b), Sprays et connexions sur les fibrés tangents d'ordre supérieur, C.R. Acad. Sc. Paris, 293 I, 51-54.

de LEÓN, M. & VÁZQUEZ ABAL, E. - (1984), On the geometry of the tangent bundle or order 2, Preprint.

de WETT, J. - (1948), Proc. Camb. Phil. Soc., 44, 546-559.

DIRAC, P.A.M. - (1964), Lectures on Quantum Mechanics, Belfer Graduate School of Sc. Monogr. Ser., nº 3, N.Y.

DODSON, C.T.J. & RADIVOIOVICI, M.S. - (1982), Tangent and frame bundles of order two, I. Ann. St. Univ. Iasi, 28.

DOMBROWSKI, P. - (1962), On the geometry of the tangent bundle, J. Reine Ang. Math., 210, 73-88.

EHRESMANN, Ch. -

(1951), Les prolongements d'une variété différentiable: 1 Calcul des Jets, prolongement principal, C.R. Acad. Sc. Paris, 233, 598-600.

(1954), Extension du calcul des Jets aux Jets non-holonomes, C.R. Acad. Sc. Paris, 239, 1762-1764.

(1955)(a), Applications de la notion de Jet non-holonome, C.R. Acad. Sc. Paris, 240, 397-399.

(1955)(b), Les prolongements d'une space fibré différentiable, C.R. Acad. Sc. Paris, 240, 1755-1757.

ELIOPOULOS, H. -
(1962), Structures presque tangentes sur les variétés différentiables, C.R. Acad. Sc. Paris, 255, 1563-1565.
(1965), On the general theory of differentiable manifolds with almost tangent structure, Canad. Math. Bull., 8, 721-748.
(1966), Structures r-tangentes sur les variétés différentiables, C.R. Acad. Sc. Paris, 263, 413-416.

ELLIS, J. - (1975), A canonical formalism for an acceleration dependent Lagrangian, J. Phys. A, 3, (4), 496-505.

FLANDERS, H. - (1963), Differential Forms with applications to the Physical Sciences, Acad. Press, N.Y.

FRANCAVIGLIA, M. & KRUPKA, D. - (1982), The Hamiltonian formalism in higher order variational problems, Ann. Inst. Henri Poincaré, 37, (3), 295-315.

FRÖLICHER, A. & NIJENHUIS, A. - (1956), Theory of vector-valued differential forms, Ind. Math., 18, 338-385.

GALLISSOT, T. -
(1952), Les formes extérieures en Mécanique, Ann. Inst. Fourier, Grenoble, 4, 145-297.
(1957), Les formes extérieures et la Mécanique des milieux continus, C.R. Acad. Sc. Paris, 244 A, 2347-2349.

GARCIA, P. -
(1968), Geometria simplética en la teoria Classica de Campos, (Doctoral thesis), Coll. Math., 19, 1-66.
(1974), The Poincaré-Cartan invariant in the Calculus of Variations, Symp. Math., 14, 219-246.

GARCIA, P. & MUÑOZ, J. - (1983), On the Geometrical structure of Higher order Variational Calculus, IUTAM-ISIMM Symposium on "Modern Developments in Analytical Mechanics", Torino, 127-147.

GARCIA, P. & RENDÓN, P. -

(1969), Symplectic approach to the theory of quantized fields, Comm. Math. Phys. 13, 24-44.

(1971), Symplectic approach II, Archiv for Rat. Mech. & Anal. 43, 101-124.

GAWEDSKY, K. - (1972), On the geometrization of the canonical formalism in the Classical Field theory, Rep. Math. Phys., 3, (4), 307-326.

GELFAND, I. & FOMIM, S. - (1963), Calculus of Variations, Prentice-Hall, Englewood Cliff.

GODBILLON, C. - (1969), Géométrie Différentielle et Mécanique Analytique, Hermann, Paris.

GOLDSCHMIDT, H. & STERNBERG, S. - (1973), The Hamilton-Jacobi formalism in the Calculus of Variations, Ann. Inst. Fourier (Grenoble), 23, 203-267.

GOLDSTEIN, H. - (1950), Classical Mechanics, Addison-Wesley, Mass.

GOLUBITSKY, M. & GUILLEMIN, U. - (1973), Stable mappings and their singularities, Springer Verlag, N.Y.

GOTAY, M. - (1979), Presymplectic manifolds, geometric constraint theory and the Dirac-Bergmann theory of constraints, (Doctoral Thesis), Univ. of Maryland, 1979, 1-198.

GOTAY, M., NESTER, J. & HINDS, G. - (1978), Presymplectic manifolds and the Dirac-Bergmann theory of constraints, J. Math. Phys., 19, (11), 2388-2399.

GOTAY, M. & NESTER, J. -

(1979)(a), Presymplectic Lagrangian Systems I: the constraint algorithm and the equivalence theorem, Ann. Inst. Henri Poincaré, 30, (2), 129-142.

(1979)(b), Presymplectic Lagrangian Systems II: the second-order equation problem. Preprint pp. 79-141, Research Paper nº 431, August, 1-24.

GRÄSSER, H. - (1970), On a general Hamilton-Jacobi theory for n^{th} order single integral Calculus of Variations, I. Publ. Inst. Lombards, Acc. Sc. Lettere A, 104, 322-355; ibid 105, 721-741.

GREEN, A. -

(1947), Self-energy and interaction-energy in Podolsky's Generalized Electrodynamics, Phys. Rev. 72, (7), 628-631.

(1948), On infinities in Generalized Meson-Field Theory, Phys. Rev. 73, 26-29.

GRIFFITHS, Ph. - (1983), Extensive Differential Systems and the Calculus of Variations, Prog. in Math., 25, Birkhäuser, 1-335.

GRIFONE, J. -

(1972)(a), Estructure presque-tangente et connexions I, Ann. Inst. Fourier, Grenoble, 22, (3), 287-334.

(1972)(b), Estructure presque-tangente et connexions II, Ann. Inst. Fourier, Grenoble, 22 (4), 291-338.

HAYES, G. - (1969), Quantization of the Generalized Hamiltonian, J. Math. Phys. 10, 1555-1558.

HAYES, G. & JANKOWSKI, J. - (1968), Quantization of Generalized Mechanics, Nuovo Cimento, 58, 494-497.

HERMANN, R.

(1968), <u>Differential Geometry and the Calculus of Variations</u>, Acad. Press, N.Y.

(1970), <u>Vector bundles in Mathematical Physics</u>, vol. I, II, Benjamin, N.Y.

(1973), <u>Geometry, Physics and Systems</u>, Marcel Dekker, N.Y.

HORÁK, M. & KOLÁR, I. - (1983), On higher order Poincaré-Cartan forms, Czechoslovak Math. J., 33, (108), 467-475.

HOUH, C. - (1969), On a Riemannian manifold M_{2n} with an almost tangent structure, Canad. Math. Bull., 12, 759-769.

HUSEMOLLER, D. - (1975), Fibre bundles, 2nd ed., Springer Verlag, N.Y.

KANAI, E. & TAKAGI, S. - (1946), Some remark's on Bopp's field theory, Prog. Theor. Phys., 1, (2), 43-55.

KATAYAMA, Y. - (1953), Theory of the interactions with higher derivatives and its applications to the non-local interaction, Prog. Theor. Phys., 10, (1), 31-56.

KAWAGUCHI, A. - (1960), The theory of problems in the Calculus of Variations whose Lagrangian functions involves 2nd order derivatives, a new approach, Ann. Mat. Pure Appl., 4, (55), 77-104.

KIMURA, T. - (1972), On the Hamiltonian formalism for general Lagrangians with higher order derivatives, Lett. Nuovo Cimento, 5, (1), 81-85.

KOBAYASHI, S. & NOMIZU, K. - (1963), (1969), The Foundations of Differential Geometry, I,II, Willey Intersc., N.Y.

KOCKINOS, C. -

(1979), Construction of the E. Cartan fundamental form I - General Theory, Tensor 33, 227-241 and II, ibid. 248-258.

(1982), On the local equivalence of vector fields with a singularity and E. Cartan fundamental form, Tensor 39, 179-186.

KOSZUL, J. - (1960), Lectures on Fibre bundles and Differential Geometry, Tata Inst., Bombay.

KLEIN, J. -

(1962), Espaces variationnels et Mécanique, Ann. Inst. Fourier, (Grenoble), 12, 1-124.

(1963)(a), Opérateurs différentiels sur les variétés presque tangentes, C.R. Acad. Sc. Paris, 257 A, 2392-2394.

(1963)(b), Les Systèmes Dinamiques abstraits, Ann. Inst. Fourier, (Grenoble), 13, 2, 191-202.

KLEIN, J. & VOUTIER, A. - (1968), Formes extérieures génératrices de sprays, Ann. Inst. Fourier, Grenoble, 18, 241-250.

KOESTLER, J. & SMITH, J. - (1965), Some Developments in Generalized Classical Mechanics, Amer. J. Phys., 33, 140-145.

KOLÁŘ, I.

(1975), On the Hamilton formalism in fibered manifolds, Scripta Fac. Sci. Nat. UJEP, Brunensis, Physica 5, 249-254.

(1982)(a), On the second tangent bundle and generalized Lie derivatives, Tensor, N.S., 38, 98-102.

(1982)(b), Lie derivatives and Higher order Lagrangians, Proc. Conf. on Diff. Geometry and its Appl., Publ. Univ. Karlova, Prague, 117-123.

(1983), Some geometric aspects of the higher order Variational Calculus, to appear in Proceed. Conf. on Diff. Geometry and its Appl., Nové Město na Moravě, Sep. 5-9, 1983.

(1984), A geometrical version of the higher order Hamiltonian Formalism in fibred manifolds, preprint.

KRUGER, J. & CALLEBAUT, D. - (1968), Comments on Generalized Mechanics, Amer. J. Phys., 36, 557-558.

KRUPKA, D. -

(1971), Lagrange theory in fibered manifolds, Rep. Math. Phys., 2, (2), 121-133.

(1973), Some Geometric aspects on Variational problems in fibred manifolds, Folia Fac. Sc. Nat. Univ. Brunensis, XIV, (10), 1-65.

(1974)(a), On generalized invariant transformations, Rep. Math. Phys., 5, (3), 355-360.

(1974)(b), A setting for Generally invariant Lagrangian structures in tensor bundles, Bull. Acad. Pol. Sc. Ser. Math. Phys., XXII, (9), 967-972.

(1974)(c), On the structure of the Euler-Mapping, Arch. Math., Scrip. Fac. Sc. Nat., UJEP Brumensis, 10, 55-62.

(1975)(a), A geometric theory of ordinary first order variational problems in fibered manifolds,I: Critical sections, J. Math., Anal. Appl., 49, (1), 180-206.

(1975)(b), A geometric theory of ordinary first order variational problems in fibered manifold,II: Invariance, J. Math. Anal. Appl, 49, (2), 469-475.

(1977), A map associated to the Lagrangian forms on the Calculus of Variations in fibered manifolds, Czech. Math. J., 27, (102), 114-118.

KRUPKA, D. & TRAUTMAN, A. - (1974), General invariance of Lagrangian structures, Bull. Acad. Pol. Sc. Serie Math. Phys., XXII, (2), 207-211.

LANG, S. - (1972), Differential manifolds, Addison Wesley, Mass.

LEECH, J. - (1965), Classical Mechanics, Math., London.

LEHMANN-LEJEUNE, J. -

(1964), Sur l'intégrabilité des certaines G-structures, C.R. Acad. Sc. Paris, 258, 5326-5329.

(1966), Intégrabilité des G-structures définies par une 1-forme 0-déformable à valeurs dans le fibré tangent, Ann. Inst. Fourier, Grenoble, 16, (2), 329-387.

LIBERMANN, P. -

(1963), On sprays and higher order connections, Proc. Nat. Acad. Sc. USA, 49, 459-462.

(1967), Connexions d'ordre supérieur et tenseur de structure, Atti Conv. Intern. Geom. Differ., Bologna.

LICHNEROWICZ, A. - (1975), Variété symplectique et dynamique associée à une sous-variété, C.R. Acad. Sc. Paris, 280 A, 523-527.

MANGIAROTTI, L. & MODUGNO, M. - (1983), Some results on the Calculus of Variations on jets spaces, Ann. Inst. Henri Poincaré, 39, (1), 29-43.

MARTINET, J. - (1970), Singularités des formes différentielles, Ann. Inst. Fourier, Grenoble, 20, (1), 95-178.

MATTHEWS, P. - (1949), A note on Podolsky Electrodynamics, Proc. Camb. Phil. Soc., 45, 44-451.

MICHOR, P.W. - (1980), Manifolds of Differentiable Mappings, Shiva Math. Series, nº 3, Shiva Publ.,

MIMURA, F. - (1974), Generalized Formalism of Mechanics, Bull. Kyushu Inst. Tech., 21, 1-20.

MONTGOMERY, D. - (1946), Relativistic interaction of electrons on Podolsky's Generalized Eletrodynamics, Phys. Rev., 69, 117-124.

MORIMOTO, A. -

(1969), Prolongations of geometric structures, Lect. Notes, Math. Inst. Nagoya Univ.

(1970), Liftings of some types of tensor fields and connections to tangent bundles of p^r-velocities, Nagoya Math. J., 40, 13-31.

MUÑOZ, J. -

(1983)(a), Canonical Cartan equations for higher order variational problems, to be publihsed in J. Geom. and Phys.

(1983)(b), Pre-symplectic structure for higher order variational problems, to be published in "Proceedings Conf. on Diff. Geometry and Appl., Checoslovaquia, 1983.

(1983)(c), Teoria de Hamilton-Cartan para los problemas Variacionales de ordem superior sobre variedades fibradas, (Doctoral Thesis), Univ. Salamanca, Spain.

(1984), Formes de structure et transformations infinitésimales de contact d'ordre supérieure, C.R. Acad. Sc. Paris, 298, Ser. I, (8), 185-188.

MUSICKY, D. -
(1978)(a), On the canonical formalism in field theory with derivatives of higher order: canonical transformations, J. Phys. A, 11, (1), 39-53.
(1978)(b), On canonical formalism with derivatives of higher order, Publ. Inst. Math. (Beograd), 23, (37), 141-153.

OSTROGRADSKY, M. - (1850), Mémoire sur les équations différentielles relatives aux problèmes des isopérimetres, Mém. Acad. Sc. St. Petersburg, 6, 385-517.

OLIVA, W. - (1970), Lagrangian systems on manifolds, Celest. Mech., 1, 491-511.

PAIS, A. & UHLENBECK, G. - (1950), On Field theories with non-localized action, Phys. Rev. D, 79, (1), 145-165.

PEETRE, J. - (1978), The Euler derivative, Math. Scand., 42, 313-333.

PODOLSKI, B. - (1942), A Generalized Electrodynamics I, Non-quantum, Phys. Rev. 62, 68-71.

PODOLSKI, B. & KIKUCHI, C. - (1944), A Generalized Electrodynamics, II - Quantum, Phys. Rev. 65, 228-235.

PODOLSKI, B. & SCHWED, P. - (1948), Review of a Generalized Electrodynamics, Rev. Mod. Phys., 20, 40-50.

POMMARET, J.F. - (1978), Systems of partial differential equations and Lie pseudogroups, Gordon and Breach, N.Y., Paris.

POOR, W. - (1981), Differential Geometric structures, Mc Graw-Hill, N.Y.

PNEVMATIKOS, S. - (1983), Singularités en Géométrie Symplectique, in Symplectic Geometry, Research N. in Math. 80, Ed. A. Crumeyrolle & J. Grifone, Pitman Books, London, 184-216.

RAYAN, C. - (1972), On Lagrangians with higher order derivatives, Amer. J. Phys., 40, 386-390.

RIEWE, F. - (1972), Relativistic Classical spinning particle Mechanics, Nuovo Cimento, 8, (1), 271-277.

RIHAI, F. - (1972), On Lagrangians with higher order derivatives, Amer. J. Phys., 40, 386-390.

RODRIGUES, L. & RODRIGUES, P.R. - (1970), Futher developments in Generalized Classical Mechanics, Amer. J. Phys., 38, (5), 557-560.

RODRIGUES, P.R. -

(1975), Sur les systèmes Mécaniques Lagrangiens homogènes d'ordre supérieur, C.R. Acad. Sc. Paris, 281, 643-646.

(1976), Sur les systèmes Mécaniques Généralisés, C.R. Acad. Sc. Paris, 282, 1307-1309.

(1977), On generating forms of k-generalized Lagrangian and Hamiltonian systems, J. Math. Phys., 18, (9), 1720-1723.

(1984)(a), On Lagrangian equations with generic constraints, Ann. Acad. Bras. Ciênc. 56, (1), 13-16.

(1984)(b), Mecânica em Fibrados dos Jatos de Ordem Superior, (Mechanics on higher order Jet bundles), Publ. Dep. Geometrya y Topologia, Univ. Santiago de Compostela, 1-68.

(1985), On the canonical form of higher order Lagrangians, Preprint.

ROUX, A. - (1970), Jets et connexions, Publ. Math. Univ. Lyon, 7, (4), 1-42.

RUND, H. - (1966), The Hamilton-Jacobi theory in the Calculus of Variations, Van Nostrand, N.Y.

SHADWICK, W. - (1982), The Hamiltonian formulation of regular r^{th} order Lagrangian field theories, Lett. Math. Phys., 6, 409-416.

SHANMUGADHASAN, S. -
(1963), Generalized Canonical formalism for degenerate dynamical systems, Proc. Camb. Phil. Soc., 59, 743.
(1973), Canonical formalism for degenerate Lagrangians, J. Math. Phys., 14, 677.

STEENROD, N. - (1951), <u>The topology of fibre bundles</u>, Princeton Univ. Press, Princeton, N.J.

TAKASU, T. - (1968), Various Hamiltons canonical formalisms as non-connection method for various connection geometries in the large, Yokohama Math. J., 16, (1), 15-62.

TANIUTI, T. -
(1955), On the theories of higher derivatives and non-local couplings, I, Prog. Theor. Phys., 13, (5), 505-521.
(1956), On the theories of higher derivatives and non-local couplings II, Progr. Theor. Phys., 15, (1), 19-36.

THIELHEIM, K. - (1967), Note on Classical fields of higher order, Proc. Camb. Phyl. Soc. 91, 798-803.

TRAUTMAN, A. - (1967), Noether equations and conservation Laws, Comm. Math. Phys., 6, 248-261.

TULCZYJEW, W. -
(1975)(a), Sur la différentielle de Lagrange, C.R. Acad. Sc. Paris, 280 A, 1295-1298.
(1975)(b), Les jets généralisés, C.R. Acad. Sc. Paris, 281 A, 349-352.
(1977), The Lagrange complex, Bull. Soc. Math. France, 105, 419-431.

Ver EECKE, P. - (1967), <u>Calcul des Jets</u>, Publ. Soc. Mat. S. Paulo, SP.

VILMS, J. - (1967), Connections on tangent bundles, J. Diff. Geometry, 1, 235-243.

VINOGRADOV, A. & KUPERSHMIDT, B. - (1977), The structure of Hamiltonian Mechanics, Math. Surveys, 32, (14), 177-243.

WALKER, A. - (1961), Almost-product structures, Proc. Symp. Pure Math. III, 94-100.

WILLMORE, T. - (1956), Parallel distributions on manifolds, Proc. London Math. Soc. B, (6), 191-204.

WHITTAKER, E.T. - (1959), _A treatise on the Analytical Dynamics of Particle and Rigid Bodies_, Ch. X, p. 265, Camb. Univ. Press, Cambridge.

YANO, K. & DAVIES, E. - (1975), Differential Geometry on Almost Tangent Manifolds, Ann. Mat. Pure Appl., 4, (103), 131-160.

YANO, K. & ISHIHARA, S. - (1973), Tangent and Cotangent bundles; Differential Geometry, Marcel Dekker, N.Y.

SUBJECT INDEX